Pediatric Reference Intervals

Seventh Edition

Pediatric Reference Intervals
Seventh Edition

Edited by

Steven J. Soldin, PhD
Director, Bioanalytical Core Laboratory
Georgetown-Howard Universities
Center for Clinical and Translational Sciences
Georgetown University
Washington, DC

Edward C. Wong, MD
Division of Laboratory Medicine
Children's National Medical Center
George Washington University
School of Medicine and Health Sciences
Washington, DC

Carlo Brugnara, MD
Department of Laboratory Medicine
Children's Hospital
Harvard University
Boston, MA

Offie P. Soldin, PhD, MBA
Departments of Oncology and Medicine
Georgetown University School of Medicine
Washington, DC

Editor Emeritus
Jocelyn M. Hicks, PhD
JMBH Associates Health Care Management Consultants
Washington, DC

Former Editor
Kurt C. Gunter, MD
Regenerative Medicine
Hospira Inc.
Lake Forest, IL

Major contributions from
Alain Colaço
and
the Colaço Fellows

1850 K Street, NW, Suite 625
Washington, DC 20006-2213

©2011 American Association for Clinical Chemistry, Inc. All rights reserved. No part of this publication may be reproduced, stored in a retrieval system, or transmitted in any form by electronic, mechanical, photocopying, or any other means without written permission of the publisher.

1 2 3 4 5 6 7 8 9 0   PCP   13 12 11

Printed in the United States of America

Library of Congress Cataloging-in-Publication Data

Pediatric reference intervals / edited by Steven J. Soldin ... [et al.] ; editor emeritus, Jocelyn M. Hicks; former editor, Kurt C. Gunter ; major contributions from Alain Colaço and the Colaço Fellows. – 7th ed.
    p. ; cm.
  Includes bibliographical references and index.
  ISBN 978-1-59425-101-6 (alk. paper)
  1. Children–Diseases–Diagnosis. 2. Diagnosis, Laboratory. 3. Clinical chemistry. 4. Reference values (Medicine) I. Soldin, Steven J. II. American Association for Clinical Chemistry.
  [DNLM: 1. Clinical Chemistry Tests. 2. Reference Values. 3. Adolescent. 4. Child. 5. Infant. 6. Laboratory Techniques and Procedures. QY 16]
  RJ51.L3P455 2011
  618.92'0075–dc23
                                            2011014950

We dedicate this book to our children and grandchildren. We hope that children worldwide will benefit from this seventh edition of Pediatric Reference Intervals (*formerly* Pediatric Reference Ranges)

CONTENTS

Foreword	xi
Preface	xiii
Introduction	xvii

Chemistry Tests

Adiponectin	3
Alanine Aminotransferase (ALT, SGPT)	4
Albumin	6
Aldolase	8
Aldosterone	9
Alkaline Phosphatase (ALP)	11
Alpha$_1$-Antitrypsin (α_1-AT)	14
Alpha-Fetoprotein (AFP)	15
Amino Acids (Plasma)	17
Alanine	17
β-Alanine	17
Anserine	17
α-Aminoadipic Acid	17
α-Amino-n-butyric Acid	18
γ-Aminobutyric Acid	18
β-Aminoisobutyric Acid	18
Arginine	18
Asparagine	19
Aspartic Acid	19
Carnosine	19
Citrulline	19
Cystathionine	20
Cystine	20
Ethanolamine	20
Glutamic Acid	20
Glutamine	21
Glycine	21
Histidine	21
Homocystine	21
Hydroxylysine	22
Hydroxyproline	22
Isoleucine	22
Leucine	22
Lysine	23
Methionine	23
1-Methylhistidine	23
3-Methylhistidine	23
Ornithine	24
Phenylalanine	24
Phosphoethanolamine	24
Phosphoserine	24
Proline	25
Sarcosine	25
Serine	25
Taurine	25
Threonine	26
Tryptophan	26
Tyrosine	26
Valine	26
Amino Acids (Urine)	27
Alanine	27
β-Alanine	27
Anserine	27
α-Aminoadipic Acid	27
α-Amino-n-butyric Acid	27
γ-Aminobutyric Acid	27
β-Aminoisobutyric Acid	27
Arginine	27
Asparagine	28
Aspartic Acid	28
Carnosine	28
Citrulline	28
Cystathionine	28
Cystine	28
Ethanolamine	28
Glutamic Acid	28
Glutamine	29
Glycine	29
Histidine	29
Homocystine	29
Hydroxylysine	29
Hydroxyproline	29
Isoleucine	29

Leucine	29	Complement Fraction — C_5	75
Lysine	30	Complement Fraction — C_{5a}	75
Methionine	30	Complement Fraction — C_7	76
1-Methylhistidine	30	Complement Fraction — Factor D	76
3-Methylhistidine	30	Complement Fraction — Factor H	77
Ornithine	30	Complement Fraction — Factor I	77
Phenylalanine	30	Complement Fraction — Properdin	78
Phosphoethanolamine	30	Copper	79
Phosphoserine	30	Cortisol	80
Proline	31	C-Reactive Protein (CRP)	82
Sarcosine	31	Creatine Kinase	83
Serine	31	Creatine Kinase Isoenzymes (CPK Isoenzymes)	85
Taurine	31	Creatinine	86
Threonine	31	Creatinine (Urine)	88
Tryptophan	31	Cystatin C	89
Tyrosine	31	Dehydroepiandrosterone (DHEA)	90
Valine	31	Dehydroepiandrosterone Sulfate (DHEAS)	92
Ammonia	32	Deoxycorticosterone (DOC)	95
Amylase	33	11-Deoxycortisol (Compound S)	96
Androstenedione	35	11-Deoxycortisol (Compound S, Metyrapone Test)	97
Apolipoprotein A-1	38	Dihydrotestosterone (DHT)	98
Apolipoprotein (a)	40	Dopamine (Urine)	99
Apolipoprotein B	41	Epinephrine/Adrenaline (Plasma)	100
Apolipoprotein CII	43	Epinephrine/Adrenaline (Urine)	101
Apolipoprotein CIII	43	Erythropoietin	102
Apolipoprotein E	43	Estradiol	103
Aspartate Aminotransferase (AST, SGOT)	44	Estrone	106
Base Excess	46	Ferritin	107
Bicarbonate (HCO_3^-)	46	Folic Acid	109
Bile Acids (Total) (Total 3-α-Hydroxy Bile Acids)	47	Follicle Stimulating Hormone (FSH)	110
Bilirubin (Conjugated)	48	Free Fatty Acids	112
Bilirubin (Total)	49	Fructosamine	113
Brain Natriuretic Peptide	51	Gamma-Glutamyltransferase (GGT)	114
(N-Terminal Pro)-Brain Natriuretic Peptide	52	Gastrin	116
C-Peptide	53	Ghrelin	116
C1 Esterase Inhibitor	54	Globulins Total, Calculated	117
Calcium	55	Glucagon	118
Calcium, Ionized	58	Glucose	119
Carbon Dioxide (CO_2)	59	Glucose (Urine)	120
Carbon Dioxide, Partial Pressure (pCO_2)	61	Glucose Cerebrospinal Fluid (CSF Glucose)	120
Carnitine (Total)	62	Glutathione Peroxidase Activity	121
Ceruloplasmin	63	Growth Hormone	122
Chloride	64	Growth Hormone (Urine)	123
Cholesterol	65	Haptoglobin	124
Coenzyme Q10	68	Hemoglobin A_{1c} (HbA_{1c})	125
Complement Fraction — C_{1r}	69	High-Density Lipoprotein Cholesterol (HDL-C)	126
Complement Fraction — C_2	69	Homocysteine (Total)	128
Complement Fraction — C_{3a}	70		
Complement Fraction — C_{3c}	71		
Complement Fraction — C_4	73		

Homovanillic Acid (HVA; 4-Hydroxy-3-Methoxyphenylacetic Acid) (Urine)	129	Potassium	182
		Prealbumin (Transthyretin)	184
Hyaluronic Acid	130	Pregnenolone	186
β-Hydroxybutyrate/3-Hydroxybutyrate	130	17-OH-Pregnenolone	186
5-Hydroxyindoleacetic Acid (5HIAA) (Urine)	131	Progesterone	187
		Prolactin	189
17α-Hydroxyprogesterone (17α-OHP)	132	Protein, Total	191
Immunoglobulin A (IgA)	134	Protein Cerebrospinal Fluid (CSF Protein)	193
Immunoglobulin D (IgD)	135	Pyruvate	194
Immunoglobulin E (IgE)	136	Renin (Plasma)	194
Immunoglobulin G (IgG)	138	Renin Activity	195
Immunoglobulin G Subclass (IgG-Subclass)	140	Retinol Binding Protein (RBP)	196
		Reverse Triiodothyronine (rT_3)	197
Immunoglobulin Light Chains (Kappa and Lambda)	141	Selenium	197
		Serum Osteocalcin	199
Immunoglobulin M (IgM)	142	Sex Hormone Binding Globulin (SHBG)	200
Insulin (See C-Peptide on page 53)	144	Sodium	201
		Sweat Electrolytes	202
Insulin-Like Growth Factor Binding Protein-3 (IGF Binding Protein-3, IGF Bp-3)	146	Testosterone	203
		Thyroid Stimulating Hormone (TSH)	206
		Thyroxine (T_4)	210
Insulin-Like Growth Factor-I (IGF-I), Somatomedin-C	148	Thyroxine Binding Globulin (TBG)	213
		Thyroxine, Free (Free T_4)	214
Insulin-Like Growth Factor-II (IGF-II)	150	Transferrin	218
Iodide (Urine), (UI)	151	Transferrin Saturation	220
Iron	152	Triglycerides	221
Iron-Binding Capacity, Total (TIBC)	154	Triiodothyronine (T_3)	224
Lactate	156	Triiodothyronine, Free (Free T_3)	227
Lactate Dehydrogenase (LDH)	157	Triiodothyronine Uptake Test (T_3U)	229
Lactate/Pyruvate Ratio	159	Troponin I	231
Lead	159	Urea Nitrogen	232
Leptin	160	Uric Acid	234
Lipase	161	Urine Volume (24 h)	236
Low-Density Lipoprotein-Cholesterol (LDL-C)	163	Vanillylmandelic Acid (VMA; 4-Hydroxy-3-Methoxymandelic Acid) (Urine)	237
Luteinizing Hormone (LH)	165	Vitamin A (Retinol)	238
Magnesium	167	Vitamin B_{12}	239
Manganese	169	25-Hydroxy Vitamin D_3 (25 OH VIT D_3)	240
Metanephrine (Urine)	169	Vitamin E (α-Tocopherol)	242
Methylmalonic Acid	170	Zinc	243
$β_2$-Microglobulin	170	Zinc Protoporphyrin	245
Mucopolysaccharides (Urine)	171		
Norepinephrine/Noradrenaline (Plasma)	172	***Hematology Tests***	
Norepinephrine/Noradrenaline (Urine)	173	Atypical Lymphocyte Count (Relative)	249
Normetanephrine (Urine)	173	Basophil Count (Absolute)	250
Osmolality	174	Basophil Count (Relative)	251
Oxygen, Partial Pressure (pO_2)	175	Cellular Hemoglobin Concentration Distribution Width (HDW)	253
Oxygen Saturation	176		
Parathyroid Hormone (PTH)	177	Eosinophil Count (Absolute)	254
Peptide YY (PYY)	178	Eosinophil Count (Relative)	256
pH	179	Hematocrit	258
Phosphorus	180		

Hematopoietic Cell Progenitor (Absolute)	260	Neutrophil Count (Absolute)	285
Hemoglobin	261	Neutrophil Count (Relative)	287
Hemoglobin A	263	Nucleated Red Blood Cell Count (Absolute)	288
Hemoglobin A2	264	Nucleated Red Blood Cell Count (Relative)	289
Hemoglobin F	265	Platelet Count	290
Immature Granulocytes (Absolute)	266	Red Cell Count	292
Immature Granulocytes (Relative)	267	Red Cell Distribution Width, CV (RDW)	294
Immature Platelet Fraction (IPF)	268	Red Cell Distribution Width, SD (RDW)	296
Immature Reticulocyte Fraction (IRF)	269	Reticulocyte Cellular Hemoglobin Content (CHR)	297
Lymphocyte Count (Absolute)	270	Reticulocyte Count (Absolute)	298
Lymphocyte Count (Relative)	272	Reticulocyte Count (Relative)	299
Mean Corpuscular Hemoglobin (MCH)	274	Reticulocyte Hemoglobin Equivalent (Ret-He)	301
Mean Corpuscular Hemoglobin Concentration (MCHC)	276	Soluble Transferrin Receptor (sTfR)	302
Mean Corpuscular Volume (MCV)	278	White Cell Count	303
Mean Platelet Volume (MPV)	280		
Monocyte Count (Absolute)	282		
Monocyte Count (Relative)	283		

FOREWORD

from *Pediatric Reference Ranges,* 4th Edition

This invaluable volume on pediatric laboratory medicine is a very important contribution to patient care, now made even more useful by the addition of a greatly expanded chemistry section and more complete hematology reference ranges. Every children's hospital, pediatric department, and clinical laboratory should have this well-designed and easy-to-read manual close athand. The editors are to be commended for their splendid addition to the quality of laboratory diagnosis.

David G. Nathan, MD
Dana Farber Cancer Institute
Boston, Massachusetts

PREFACE

Children are unique individuals and not miniature adults. For example, neonates and premature infants have immature hepatic, renal, and pulmonary function. This affects the way these individuals handle drugs. Furthermore, their pediatric reference intervals for numerous analytes in many cases are clearly different from those found in older children and adults.

How Drug Metabolism Changes with Age

The majority of drugs are metabolized by the hepatic intestinal microsomal enzyme systems, which are under genetic control and subject to many factors that influence its activity. If, for instance, patients eat charbroiled meat or inhale the smoke of cigarettes, the activity of the microsomal system is enhanced, increasing the clearance of the drug being given and shortening its half life. Such patients may require dosage adjustments. Age is another major influence. In the neonate, the hepatic microsomal enzyme systems and renal and hepatic function are immature. Therefore, a much smaller dose per kilogram must be given to attain the same therapeutic concentration of a drug. During the first four months of life, infants need to be followed closely, because the enzyme activity is increasing, causing major changes in dosage requirement. In children from about six months of age to puberty, the hepatic microsomal enzyme system has approximately double the activity of the adult, requiring about double the dose per kilogram to achieve the same therapeutic concentration as an adult.

As the youth goes through puberty, the activity of the system begins decreasing, and eventually the individual has essentially the same hepatic microsomal enzyme system activity as an adult. One has to monitor closely if he/she is on drug therapy and going through puberty, because the activity of the enzyme system is changing a great deal over that time period.

The Concept of Reference Intervals

In order to determine if a patient suffers from a particular disease, clinicians will order tests that will confirm or refute their suspicions. A laboratory test is often the starting point for the differential diagnosis of a particular disease from among a spectrum of possible diseases. There are many different ways to establish age- and sex-related reference values, each method with its own inherent advantages and disadvantages. These include obtaining specimens from known "healthy" individuals and after removing the extreme outliers, tabulating values from the lowest to highest and removing the 0–2.5th percentile and the 97.5–100.00th percentile, thereby leaving an interval of the 2.5–97.5th percentile for any particular analyte. One can also use the mean ± standard deviation (SD) approach, but for this method to be valid, the frequency distribution of values must be Gaussian (symmetrical or bellshaped distribution). If the data are skewed and non-Gaussian, it can often be made Gaussian by plotting the log of the value instead of the value itself. Once a Gaussian distribution is obtained, one can calculate the reference interval from the mean ± 2 SD.

Fig. 1 shows normal interval plots for two different analytes. In (A) there is no overlap between the diseased and nondiseased populations. An example may be urinary vanillymandelic acid (VMA) or homovanillic acid (HVA) often ordered to confirm a diagnosis of pheochromocytoma or neuroblastoma. In (B),

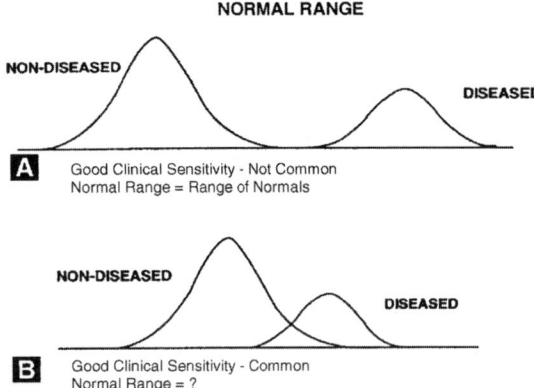

Fig. 1. Normal interval plots for two analytes.

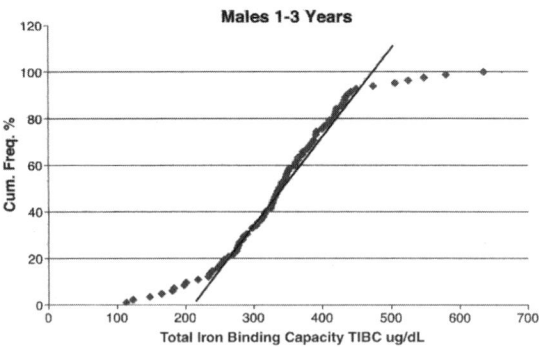

Fig. 2. Reference interval for total iron-binding capacity. *Reprinted from Clinica Chimica Acta, 342, Soldin OP, Bierbower LH, Choi JJ, Choi JJ, Thompson-Hoffman S, Soldin SJ, Serum iron, ferritin, transferring, total iron binding capacity, hs-CRP, LDL cholesterol and magnesium in children; new reference intervals using the Dade Dimension Clinical Chemistry System, pages 212–217, 2004, with permission from Elsevier.*

there is some overlap between the diseased and nondiseased populations. This situation is more common and choosing the normal reference interval cut-off point here is somewhat arbitrary. In general, most clinicians are acquainted with an interval that encompasses the 2.5–97.5th percentile of the population.

Hoffmann's Approach to Determining Reference Intervals

A third approach is to use the results obtained from hospitalized sick patients to develop the normal reference intervals. While this may seem to be a strange approach it has many advantages, especially in developing reference intervals for children. It is almost impossible to obtain a large enough population of healthy children between the ages of 1 day and 18 years who are prepared to donate blood specimens, or to obtain the informed consent necessary thereby allowing reference interval studies to be performed (1). By using either Chauvenet's or Dixon's criteria for removing outliers and then plotting % cumulative frequency versus the laboratory value (or log of the value, if non-Gaussian distribution), one obtains a straight line, which deviates from linearity at both ends. This straight line can be extended to provide the 2.5th and 97.5th percentiles for the population being studied. This approach has been described in detail by Hoffmann (2), and an example for total iron-binding capacity is shown in Fig. 2. This approach, apart from its simplicity and appeal in pediatrics, has additional advantages for analytes such as T_4 (thyroxine) and cholesterol. T_4 values are significantly increased in a "sick" population. To assess the thyroid function in the "sick" population and differentiate those patients with thyroid disease from those with other ailments, it makes eminent sense to know the T_4 reference intervals for the "sick" population. Values above or below the 2.5th–97.5th interval would then indicate the distinct possibility of thyroid disease. The same is true for cortisol, ACTH, and even cholesterol. Values of the latter decrease somewhat in a "sick population."

Reference Intervals Are Method-Dependent

It is important to emphasize that reference intervals vary with the method (technology) used to measure the analytes in question. In compiling the pediatric reference intervals for a textbook, I could not help noticing the very large differences one found for many endocrine tests, even if the same patient samples were used to establish these intervals. For example, when analyzing T_4, the male 1–30 day intervals using the Abbott IMx are 3.0–14.4 µg/dL (39–185 nmol/L), while the T_4 Gamma Coat (Baxter-Travenol Diagnostics, Cambridge, MA, USA) results are 5.9–21.5 µg/dL (76–276 nmol/L) (3). Such differences are extraordinary and indicate the considerable nonspecificity of antibodies used in immunoassays in the mid to late 1990's. One may well ask, what are we really measuring? This problem with immunoassays has led to the expanding role of tandem mass spectrometry in the clinical laboratory, especially in endocrinology where isotope dilution tandem mass spectrometry is now the state-of-the-art for measurement of steroids and thyroid hormones (T_4, T_3, and free T_4).

Children Are Not "Little Adults"

The reference intervals for many analytes vary significantly from birth through the end of puberty, as already stated. Neonates (and premature neonates), for example, have immature hepatic, renal, and pulmonary function. The normal values for blood gases, therefore, differ significantly from those of older children (pH 7.18– 7.51 versus 7.35–7.44, respectively). The same is true for electrolytes such as sodium, potassium, and calcium, where reference intervals are somewhat broader than those found later in life. For example, for ionized calcium, the reference interval for 0–1 month babies is 3.9–6.0 mg/dL (1.0–1.5 mmol/L), while in the adult it is 4.7–5.3 mg/dL (1.18–1.32 mmol/L).

Reference intervals for many enzymes vary significantly with age. For example, on the Vitros analyzer (Johnson & Johnson, Rochester, NY, USA) the male and female 97.5th percentile intervals for amylase are 30, 50, 80 and 100 U/L for 0–0.2, 0.2–0.5, 0.5–1.0, and 1–19 years, respectively. For aspartate aminotransferase (AST, SGOT), the male 2.5–97.5th percentiles using the Vitros analyzer are 30–100, 20–70, 15–45 U/L for 1–7 days, 8–30 days and 16–19 years, respectively.

Values for many endocrine tests undergo significant changes with onset of puberty. Male and female intervals can, and often do, differ significantly. In order to aid clinicians and pediatricians in interpreting laboratory data, we have recently gathered the literature available in this area, and published it in this textbook, *Pediatric Reference Intervals*. This book provides the reference intervals for a wide variety of analytes for children (males and females), from neonates to adolescents and young adults (3).

Conclusion

Anyone who uses reference intervals should realize that they are guidelines for the clinician and cannot be used as definitive indicators of health or pathologic states. Values for healthy and diseased individuals can overlap, and significant inter- and intra-individual differences exist even among healthy individuals (3).

References

1. Soldin SJ, Rifai N, Hicks JM, eds. Biochemical basis of pediatric disease, 3rd ed. Washington, DC: AACC Press, 1998.
2. Hoffmann RG. Statistics in the practice of medicine. JAMA 1963;185:864-73.
3. Soldin SJ, Wong EC, Brugnara SJ, Soldin OP, eds. Pediatric reference intervals, 7th ed. Washington DC: AACC Press, 2011.

Steven J. Soldin, PhD, FACB

INTRODUCTION

This book provides reference (normal) intervals on a wide variety of analytes for children from neonates to adolescents and young adults. We have put the emphasis on a user-friendly format. Each page has the same layout and provides information on the reference to the study. Whenever possible, the work is our own or from recent literature. The type of fluid analyzed is given along with the methodology. We have tried to include at least two references for each analyte. The names of the tests are reported alphabetically, and alternate names are sometimes given. In our studies, wherever possible, we have used a large number (n) of children in each age group, in compliance with the IFCC-recommended n of >120. The reference values are reported by age and sex, and wherever possible we have included SI units so that this book can be used throughout the world. We have included the statistical basis by which the reference intervals were calculated and the population source.

We have purposely not burdened this book with clinical information that can be found elsewhere, especially in our previous book entitled *The Biochemical Basis of Pediatric Disease,* which I co-edited with Dr. Jocelyn Hicks and Dr. Nader Rifai. It is our belief that, for the most part, when a physician or laboratory scientist seeks information about reference intervals, that is all he or she is interested in!

Anyone who uses reference intervals should realize that they are guidelines for the clinician and cannot be used as definitive indicators of health or pathological states. Values for healthy and diseased individuals often overlap, and, more importantly, considerable variations exist even among healthy individuals. Values also can vary according to the methodology used.

The first edition of this book was published, in 1995, as *Pediatric Reference Ranges*, and was co-edited by me and Dr. Jocelyn Hicks. Dr. Hicks remained as co-editor for the first three editions. Over the subsequent years, several additional editors have made significant contributions. I would like to take this opportunity to thank all of the editors for their sterling support and help. The Seventh Edition is my final version and spans approximately three decades of work in the field, which started with evaluation of pediatric reference intervals for urinary catecholamines and their metabolites while I was on faculty at The Hospital for Sick Children in Toronto, Canada. This latest edition includes data on several new chemistry platforms as well as tandem mass spectrometry data, and of course Dr. Offie P. Soldin is now included as a co-editor.

Over the years, we have also been blessed with the support of the Colaço fellows. In 1992, Alain Colaço was awarded a fellowship to work with me on this important mission, and his untimely death that year led to the annual "Alain Colaço Endowment Fund Summer Internship" award, which was made possible by a generous donation from Alain's parents in his memory. In the intervening years this endowment has led to the publication of many manuscripts in the peer-reviewed literature and to the training of approximately 40 young students in clinical research. The joy, energy, and superb intellect these students brought to their "summer job" was one of the important highlights of my career.

Finally, a heartfelt thanks to Joanna L. Grimes, Managing Editor, AACC Press, who has worked with me on several editions of the book, making key contributions to each.

Steven J. Soldin

CHEMISTRY TESTS

ADIPONECTIN				
	Male		**Female**	
Age	**n**	**mg/L**	**n**	**mg/L**
Maternal serum	360	0.1–18.1	353	0.4–18.0
Cord blood	361	3.3–59.7	348	5.1–60.7
Specimen Type(s)	Serum			
Reference(s)	Weyermann M, Beermann C, Brenner H, Rothenbacher D. Adiponectin and leptin in maternal serum, cord blood, and breast milk. Clin Chem 2006;52:2095–102.			
Method(s)	Adiponectin was measured with a commercially available ELISA (Rand D Systems, Minneapolis, MN).			
Comment(s)	Values are 2.5–97.5th percentiles.			

ALANINE AMINOTRANSFERASE (ALT, SGPT)

Test	Age	Male n	Male U/L	Female n	Female U/L
1	1–7 d	109	6–40	84	7–40
	8–30 d	168	10–40	71	8–32
	1–3 mo	178	13–39	173	12–47
	4–6 mo	135	12–42	59	12–37
	7–12 mo	130	13–45	107	12–41
2	1–3 y	50*	5–45	50*	5–45
	4–6 y	40*	10–25	40*	10–25
	7–9 y	80*	10–35	80*	10–35
	10–11 y	27	10–35	34	10–30
	12–13 y	31	10–55	49	10–30
	14–15 y	26	10–45	52	5–30
	16–19 y	40	10–40	61	5–35
3	1–30 d	50	1–25	51	2–25
	31–365 d	91	4–35	76	3–30
	1–3 y	119	5–30	115	5–30
	4–6 y	114	5–20	101	5–25
	7–9 y	102	5–25	109	5–25
	10–18 y	280	5–30	269	5–20
4	1–7 d	**	20–54	**	21–54
	8–30 d		24–54		22–46
	1–3 mo		27–54		26–61
	4–6 mo		26–55		26–51
	7–12 mo		26–59		26–55
	1–3 y		19–59		24–59
	4–6 y		24–49		24–49
	10–11 y		24–49		24–44
	12–13 y		24–68		24–44
	14–15 y		24–59		19–44
	16–19 y		24–54		19–49

Test	Age	Male and Female n	Male and Female U/L
5	0–12 mo	281	8.7–39.0

Age	Male n	Male U/L	Female n	Female U/L
6–10 y	153	9.0–68.6	332	5.9–37.0

Specimen Type(s)	1–5	Plasma/serum
Reference(s)	1	Soldin SJ, Savwoir TV, Guo Y. Pediatric reference ranges for alkaline phosphatase, aspartate aminotransferase, and alanine aminotransferase in children less than 1 year old on the Vitros 500. Clin Chem 1997;43:S199. (Abstract)
	2	Lockitch G, Halstead AC, Albersheim S, et al. Age- and sex-specific pediatric reference intervals for biochemistry analytes as measured with the Ektachem-700 analyzer. Clin Chem 1988;34:1622–5.
	3	Soldin SJ, Bailey J, Bjorn S, et al. Pediatric reference ranges for ALT. Clin Chem 1995;41:S92–3. (Abstract).
	4	Ghoshal AK, Soldin SJ. Evaluation of the Dade Behring Dimension RxL: integrated chemistry system-pediatric reference ranges. Clin Chim Acta 2003;331:135–46.
	5	Chan MK, Seiden-Long I, Aytekin M, et al. Canadian Laboratory Initiative on Reference Interval Database (CALIPER): pediatric reference intervals for an integrated clinical chemistry and immunoassay analyzer, Abbott ARCHITECT ci8200. Clin Biochem 2009;42:885–91.
Method(s)	1, 2	Vitros 500 (1) and 700 (2) (Ortho-Clinical Diagnostics, Raritan, NJ).
	3	Measured on the Hitachi 747 using Boehringer Mannheim reagents. (Boehringer Mannheim Diagnostics, Indianapolis, IN).
	4	Alanine aminotransferase (ALT) catalyzes the transamination of L-alanine to α-ketoglutarate (α-KG), forming L-glutamate and pyruvate. The pyruvate formed is reduced to lactate by lactate dehydrogenase (LDH) with simultaneous oxidation of reduced nicotinamide-adenine dinucleotide (NADH). Siemens RxL Analyzer (Siemens Healthcare Diagnostics, Deerfield, IL).
	5	Abbott Architect ci8200 (Abbott Diagnostics, Abbott Park, IL)
Comment(s)	1,3	Study used hospitalized patients and a computerized approach adapted from the Hoffmann technique to obtain the 2.5–97.5th percentiles.
	2	The study population was healthy children. Non-parametric methods were used to determine the 0.025 and 0.975 fractiles.
		*No significant differences were found for males and females. These ranges were therefore derived from combined data.
	4	**Reference ranges were obtained by comparing results from previously published data and using regression equations.
	5	1459 serum/plasma specimens from children attending select outpatient clinics were collected. Values are 2.5–97.5th percentiles.

ALBUMIN

Test	Age	Male n	Male g/dL	Male g/L	Female n	Female g/dL	Female g/L
1	1–7 d	161	2.3–3.8	23–38	132	1.8–3.9	18–39
	8–30 d	252	2.0–4.5	20–45	124	1.8–4.4	18–44
	31–90 d	199	2.0–4.8	20–48	178	1.9–4.2	19–42
	91–180 d	135	2.1–4.9	21–49	121	2.2–4.4	22–44
	181 d–1 y	295	2.1–4.7	21–47	267	2.2–4.7	22–47
2	0–5 d (<2.5 kg)	30	2.0–3.6	20–36	30	2.0–3.6	20–36
	0–5 d (>2.5 kg)	93	2.6–3.6	26–36	93	2.6–3.6	26–36
	1–3 y	50	3.4–4.2	34–42	50	3.4–4.2	34–42
	4–6 y	38	3.5–5.2	35–52	38	3.5–5.2	35–52
	7–9 y	74	3.7–5.6	37–56	74	3.7–5.6	37–56
	10–19 y	332	3.7–5.6	37–56	332	3.7–5.6	37–56
3	1–30 d	73	2.6–4.1	26–41	51	2.7–4.3	27–43
	31–182 d	58	2.8–4.6	28–46	30	2.9–4.2	29–42
	183–365 d	29	2.8–4.8	28–48	42	3.3–4.8	33–48
	1–18 y	652	3.2–4.7	32–47	626	2.9–4.2	29–42
4	1–7 d	*	2.4–3.9	24–39	*	1.9–4.0	19–40
	8–30 d		2.1–4.5	21–45		1.9–4.4	19–44
	31–90 d		2.1–4.8	21–48		2.0–4.2	20–42
	91–180 d		2.2–4.9	22–49		2.3–4.4	23–44
	181 d–1 y		2.2–4.7	22–47		2.3–4.7	23–47
	1–3 y		3.5–4.2	35–42		3.5–4.7	35–47
	4–6 y		3.6–5.2	36–52		3.6–5.2	36–52
	7–9 y		3.8–5.6	38–56		3.8–5.6	38–56
	10–19 y		3.8–5.6	38–56		3.8–5.6	38–56

Test	Age	n	Male and Female g/dL	Male and Female g/L
5	0–12 mo	129	2.8–4.7	28–47
	1–5 y	155	3.5–4.7	35–47
	6–10 y	134	3.6–4.7	36–47
	11–14 y	150	3.7–4.8	37–48
	15–20 y	182	3.4–4.9	34–49

Specimen Type(s)	1–4,5	Plasma/serum
Reference(s)	1	Soldin SJ, Morse AS. Pediatric reference ranges for albumin and total protein in children <1 year old using the Vitros 500 analyzer. Clin Chem 1998;44:A15. (Abstract)
	2	Lockitch G, Halstead AC, Albersheim S, et al. Age- and sex-specific pediatric reference intervals for biochemistry analytes as measured with the Ektachem-700 analyzer. Clin Chem 1988;34:1622–5.
	3	Soldin SJ, Bjorn S, Beatey J, et al. Pediatric reference ranges for albumin, globulin and total protein on the Hitachi 747. Clin Chem 1995;41:S93. (Abstract)
	4	Ghoshal AK, Soldin SJ. Evaluation of the Dade Behring Dimension RxL: integrated chemistry system-pediatric reference ranges. Clin Chim Acta 2003;331:135–46.
	5	Chan MK, Seiden-Long I, Aytekin M, et al. Canadian Laboratory Initiative on Reference Interval Database (CALIPER): pediatric reference intervals for an integrated clinical chemistry and immunoassay analyzer, Abbott ARCHITECT ci8200. Clin Biochem 2009;42:885–91.
Method(s)	1	Bromocresol green method. Vitros 500 (Ortho-Clinical Diagnostics, Raritan, NJ).
	2	Bromocresol green method. Vitros 700 (Ortho-Clinical Diagnostics, Raritan, NJ).
	3	Boehringer Mannheim albumin reagent (bromocresol green). Albumin was measured on the Hitachi 747 (Boehringer Mannheim Diagnostics, Indianapolis, IN).
	4	The albumin method is an adaptation of the bromocresol purple (BCP) dye-binding method. Siemens Dimension RxL Analyzer (Siemens Healthcare Diagnostics, Deerfield, IL).
	5	Abbott Architect ci8200 (Abbott Diagnostics, Abbott Park, IL).
Comment(s)	1,3	Study used hospitalized patients and a computerized approach adapted from the Hoffmann technique. Values are 2.5–97.5th percentiles.
	2	Healthy normal children. Values are 2.5–97.5th percentiles.
	4	*Reference ranges were obtained by comparing results from previously published data and using regression equations.
	5	1459 serum/plasma specimens from children attending select outpatient clinics were collected from the five age groups noted above. Values are 2.5–97.5th percentiles.

ALDOLASE

	Male and Female	
Age	n	U/L
10–24 mo	40	3.4–11.8
25 mo–16 y	23	1.2–8.8
17–64 y	30	1.7–4.9

Specimen Type(s)	Serum
Reference(s)	Visnapuu LA, Karlson LK, Dubinsky EH, et al. Pediatric reference ranges for serum aldolase. Am J Clin Pathol 1989;91:476–7.
Method(s)	Aldolase Test Stat Pak (Behring Diagnostics, La Jolla, CA).
Comment(s)	Essentially healthy children and adults were used in the study. Results are mean ±2 SDs.

ALDOSTERONE

Male and Female

Test	Age	n	ng/dL	pmol/L
1	6–9 y	25	1–24	28–666
	10–11 y	23	2–150	55–416
	12–14 y	27	1–22	28–610
	15–17 y	42	1–32	28–888
2	<1 y	*	5.8–110	160–3000
	1–<4 y		2.5–36	70–1000
	4–<10 y			
	P_1		1–22	30–600
	P_2		1.5–22	40–600
	P_3		1.5–22	40–600
	P_4		1.5–22	40–600
	P_5		1.5–22	40–600
3	Premature infants	**		
	26–28 w, day 4		5–635	139–17622
	31–35 w, day 4		19–141	527–3913
	Full-term infants			
	3 d		7–184	194–5106
	1 wk		5–175	139–4856
	1–12 mo		5–90	139–2498
	Children			
	1–2 y		7–54	194–1499
	2–10 y		3–35	83–971
	10–15 y		2–22	56–611
4	0–8 y	121	0.1–20	3–546
	>8 y	172	0.2–20	5–554

Specimen Type(s)	1,3,4	Serum
	2	Plasma/serum
Reference(s)	1	Pediatric endocrine testing. Nichols Institute, 1993:1.
	2	Soldin SJ, Rifai N, Hicks JM, eds. Biochemical basis of pediatric disease, 3rd ed. Washington, DC: AACC Press, 1988:233.
	3	Endocrinology expected values. Esoterix Endocrinology, Calabasas Hills, CA. ©2002 Esoterix, Inc., http://www.esoterix.com
	4	Soldin OP, Sharma H, Husted L, Soldin SJ. Pediatric reference intervals for aldosterone, 17alpha-hydroxyprogesterone, dehydroepoandrosterone, testosterone and 25-hydroxy vitamin D3 using tandem mass spectrometry. Clin Biochem 2009;42:823–7.
Method(s)	1	Extraction, chromatography, radioimmunoassay.
	2	See reference.
	3	Not provided.
	4	Samples were analyzed using isotope dilution liquid chromatography tandem mass spectrometry (LC/MS/MS).
Comment(s)	1	Upright posture, normal sodium diet. Results are 2.5–97.5th percentiles.
	2	*Numbers not provided. These reference ranges are for guidance only. Numbers not provided. P_1–P_5 refer to pubertal stages.
	3	**Numbers not provided. Values are based on early morning samples from subjects on ad lib sodium intake. Diurnal variations and values in pediatric patients on different sodium diets are currently unavailable.
	4	Reference intervals were determined for neonates and children 0–18 years of age. The study was conducted using outpatient samples obtained between January 1, 2004, and June 30, 2008. Values are the 2.5–97.5th percentiles.

ALKALINE PHOSPHATASE (ALP)

Test	Age	Male n	Male U/L	Female n	Female U/L
1	1–7 d	141	77–265	109	65–270
	8–30 d	203	91–375	141	65–365
	1–3 mo	251	60–360	234	80–425
	4–6 mo	129	55–325	66	80–345
	7–12 mo	113	60–300	58	60–330
2*	1–3 y	50**	129–291	50**	129–291
	4–6 y	40**	134–346	40**	134–346
	7–9 y	80**	156–386	80**	156–386
	10–11 y	27	120–488	3	116–515
	12–13 y	31	178–455	49	93–386
	14–15 y	26	116–483	52	62–209
	16–19 y	40	58–237	61	45–116
3	1–30 d	60	75–316	75	48–406
	31–365 d	132	82–383	122	124–341
	1–3 y	136	104–345	111	108–317
	4–6 y	113	93–309	113	96–297
	7–9 y	124	86–315	104	69–325
	10–12 y	111	42–362	109	51–332
	13–15 y	126	74–390	105	50–162
	16–18 y	112	52–171	110	47–119
4	1–7 d	***	121–351	***	107–357
	8–30 d		138–486		107–474
	1–3 mo		101–467		125–547
	4–6 mo		94–425		125–449
	7–12 mo		101–394		101–431
	1–3 y		185–383		185–383
	4–6 y		191–450		191–450
	7–9 y		218–499		218–499
	10–11 y		174–624		169–657
	12–13 y		245–584		141–499
	14–15 y		169–618		103–283
	16–19 y		98–317		82–169

Test	Age	Male and Female		
		n	U/L	
5	0–12 mo	93	150–507	
	1–5 y	147	152–767	
	6–10 y	126	135–537	

		Male		Female	
	Age	n	U/L	n	U/L
	11–14 y	67	92–549	91	50–415
	15–20 y	37	62–369	108	47–175

Specimen Type(s)	1,3,4,5	Plasma/serum
	2	Serum
Reference(s)	1	Soldin SJ, Savwoir TV, Guo Y. Pediatric reference ranges for alkaline phosphatase, aspartate aminotransferase, and alanine aminotransferase in children less than 1 year old on the Vitros 500. Clin Chem 1997;43:S199. (Abstract)
	2	Lockitch G, Halstead AC, Albersheim S, et al. Age- and sex-specific pediatric reference intervals for biochemistry analytes as measured with the Ektachem-700 analyzer. Clin Chem 1988;34:1622–5.
	3	Soldin SJ, Hicks JM, Bailey J, et al. Pediatric reference ranges for alkaline phosphatase on the Hitachi 747 analyzer. Clin Chem 1997;43:S198. (Abstract)
	4	Ghoshal AK, Soldin SJ. Evaluation of the Dade Behring Dimension RxL: integrated chemistry system-pediatric reference ranges. Clin Chim Acta 2003;331:135–46.
	5	Chan MK, Seiden-Long I, Aytekin M, et al. Canadian Laboratory Initiative on Reference Interval Database (CALIPER): pediatric reference intervals for an integrated clinical chemistry and immunoassay analyzer, Abbott ARCHITECT ci8200. Clin Biochem 2009;42:885–91.
Method(s)	1,2	Vitros 500 and 700 (Ortho-Clinical Diagnostics, Raritan, NJ). p-Nitrophenylphosphate.
	3	Hitachi 747. Boehringer Mannheim reagents (p-Nitrophenylphosphate). (Boehringer Mannheim Diagnostics, Indianapolis, IN).
	4	Alkaline phosphatase catalyzes the transphosphorylation of p-nitrophenylphosphate (p-NPP) to p-nitrophenol (p-NP). Followed spectrophotometrically. Siemens Dimension RxL Analyzer (Siemens Healthcare Diagnostics, Deerfield, IL).
	5	Abbott Architect ci8200 (Abbott Diagnostics, Abbott Park, IL).
Comment(s)	1,3	Study used hospitalized patients and a computerized approach adapted from the Hoffmann technique to obtain the 2.5–97.5th percentiles.
	2	The study population was healthy children. Non-parametric methods were used to determine the 0.025 and 0.975 fractiles.
		*Due to changes made by the manufacturer in slide performance, these results are lower than those published by the author.
		**No significant differences were found for males and females. These ranges were therefore derived from combined data.
	4	***Reference ranges were obtained by comparing results from previously published data and using regression equations.
	5	1459 serum/plasma specimens from children attending select outpatient clinics were collected from the five age groups noted above. Values are 2.5–97.5th percentiles.

ALPHA$_1$-ANTITRYPSIN (α_1-AT)

Test	Age	Male			Female		
		n	mg/dL	g/L	n	mg/dL	g/L
1	0–5 d	73*	92–282	0.92–2.81	73*	92–282	0.92–2.81
	1–3 y	51*	94–156	0.94–1.56	51*	94–156	0.94–1.56
	4–6 y	39*	102–157	1.02–1.57	39*	102–157	1.02–1.57
	7–9 y	39*	102–157	1.02–1.57	39*	102–157	1.02–1.57
	10–13 y	36	104–159	1.04–1.59	45	106–171	1.06–1.71
	14–19 y	46	97–203	0.97–2.03	66	113–191	1.13–1.91

Test	Age	Male and Female		
		n	mg/dL	g/L
2	0–1 mo	60	79.4–222.7	0.79–2.23
	1–6 mo	45	71.0–190.1	0.71–1.90
	6 mo–2 y	82	60.1–160.6	0.60–1.61
	2–19 y	303	70.4–178.6	0.70–1.79

Specimen Type(s)	1,2	Serum
Reference(s)		The above values have been adjusted from those published to convert results to current IFCC units using IFCC guidelines/standards.
	1	Lockitch G, Halstead AC, Quigley G, et al. Age- and sex-specific pediatric reference intervals: study design and methods illustrated by measurement of serum proteins with the Behring LN nephelometer. Clin Chem 1988;34: 1618–21.
	2	Davis ML, Austin C, Messmer BL, et al. IFCC-standardized pediatric reference intervals for 10 serum proteins using Beckman Array 360 system. Clin Biochem 1996;29:489–92.
Method(s)	1	Nephelometric with Behring LN nephelometer (Behring Diagnostics, Hoechst Canada, Inc., Montreal).
	2	Beckman Array 360 (Beckman Instruments, Brea, CA).
Comment(s)	1	Normal healthy children. This data set excluded children found to have SZ, MZ, or ZZ phenotypes by protease inhibitor typing.
		Note: Type MM, normal; Type MS, normal variant (~8% of population); MZ, heterozygous for deficiency (<2% of population); ZZ, homozygous for deficiency. PI typing necessary if result is low. Because α_1-AT is an acute phase reactant, MZ individuals may have normal values. Values are 2.5–97.5th percentiles.
		*Results from males and females combined. Values provided are 0.025–0.975 fractiles.
	2	Samples were obtained from children attending outpatient clinics. Results are 2.5–97.5th percentiles.

ALPHA-FETOPROTEIN (AFP)

Test	Age	Male		Non-Pregnant Female	
		n	ng/mL	n	ng/mL
1	0–1 mo	71	0.6–16387	58	0.6–18964
	1–12 mo	113	0.6–28.3	102	0.6–77.0
	1–3 y	134	0.6–7.9	116	0.6–11.1
	4–6 y	101	0.6–5.6	118	0.6–4.2
	7–12 y	138	0.6–3.7	120	0.6–5.6
	13–18 y	145	0.6–3.9	122	0.6–4.2

Test	Age	n	ng/mL
		Male and Female	
2	0–30 d	*	50.0–100,000
	1–3 mo		40.0–1000
	4 mo–18 y		0.0–12.0
3	At birth	260**	15,700–146,500
4	0–<3 mo	52	40.0–19953
	3–<6 mo	35	3.6–13183
	6 mo–<1 y	53	3.6–940

Age	Male		Female	
	n	ng/mL	n	ng/mL
1–<3 y	45	0.9–16.0	52	1.6–37.0

Age	n	ng/mL
	Male and Female	
3–<18 y	320	0.6–2.0

Specimen Type(s)	1	Plasma
	2–4	Serum
Reference(s)	1	Soldin SJ, Hicks JM, Godwin ID, et al. Pediatric reference ranges for alpha-fetoprotein. Clin Chem 1992;38:959. (Abstract)
	2	Dugaw KA, Jack RM, Rutledge J. Pediatric reference ranges for ferritin and AFP on the Vitros ECi analyzer. Clin Chem 2001;47:A108–9. (Abstract)
	3	Bader D, Riskin A, Vafsi O, et al. Alpha-fetoprotein in the early neonatal period—a large study and review of the literature. Clin Chim Acta 2004;349:15–23.
	4	Soldin OP, Dahlin JR, Gresham EG, et al. IMMULITE 2000 age- and sex-specific reference intervals for alpha fetoprotein, homocysteine, insulin, insulin-like growth factor-1, insulin-like growth factor binding protein-3, C-peptide, immunoglobulin E and intact parathyroid hormone. Clin Biochem 2008;41:937–42.
Method(s)	1	Abbott EIA procedure (Abbott Laboratories, Abbott Park, IL).
	2	Chemiluminescent immunoassay, Vitros ECi (Ortho-Clinical Diagnostics, Raritan, NJ).
	3	DELFIA (dissociation-enhanced lanthanide fluorescence immunoassay, PerkinElmer, Boston. MA).
	4	IMMULITE® 2000 immunoassay system (Siemens Healthcare Diagnostics, Deerfield, IL).
Comment(s)	1	Study used hospitalized patients and a computerized adaptation of the Hoffmann technique. Values are 2.5–97.5 percentiles.
	2	*Ages ranged from 1 h to 18 y with a total of 66 samples.
	3	**All patients were Caucasians. Values are 2.5–97.5 percentiles.
	4	Study was conducted at both Children's National Medical Center and Georgetown University, Washington, DC. Results were obtained from the Children's National Medical Center laboratory information system over the period January 5, 2001, to March 8, 2007. Patient results were accessed and used to establish reference intervals. All patient identifiers were removed except age and sex. Data was analyzed using the Hoffmann approach, and was computer adapted. Values are 2.5–97.5th percentiles.

AMINO ACIDS (PLASMA)

			\multicolumn{4}{c|}{Male and Female (nmol/mL)}			
Test	Age	n	Alanine	β-Alanine	Anserine	α-Aminoadipic Acid
1	Premature (first 6 wk)	*	212–504	0	—	0
	0–1 mo		131–710	0–10	0	0
	1–24 mo		143–439	0–7	0	0
	2–18 y		152–547	0–7	0	0
2	6 mo	*	182–396	—	—	—
	2 y		173–349	—	—	—
	6 y		182–319	—	—	—
	16 y		240–482	—	—	—

Specimen Type(s)	1,2	Plasma
Reference(s)	1	Shapira E, Blitzer MG, Miller JB, Africk DK, eds. Biochemical genetics: A laboratory manual. Oxford, UK: Oxford University Press, 1989;94–5.
	2	Lepage N, McDonald N, Dallaire L, et al. Age-specific distribution of plasma amino acid concentrations in a healthy pediatric population. Clin Chem 1997;43:2397–402.
Method(s)	1	Beckman amino acid analyzer. Beckman 6300 (Beckman Instruments Inc., Palo Alto, CA).
	2	Beckman amino acid analyzer. Beckman 7300 (Beckman Instruments Inc., Palo Alto, CA).
Comment(s)	1	Results are mean ±2 SDs.
	2	Results are 10–90th percentiles in healthy children.
	*See references for numbers.	

AMINO ACIDS (PLASMA)

			Male and Female (nmol/mL)			
Test	Age	n	α-Amino-n-butyric Acid	γ-Amino-butyric Acid	β-Amino-isobutyric Acid	Arginine
1	Premature (first 6 wk)	*	14–52	0	0	34–96
	0–1 mo		8–24	0–2	0	6–140
	1–24 mo		3–26	0	0	12–133
	2–18 y		4–31	0	0	10–140
2	6 mo	*	—	—	—	43–120
	2 y		—	—	—	46–90
	6 y		—	—	—	50–99
	16 y		—	—	—	68–128

Specimen Type(s)	1,2	Plasma
Reference(s)	1	Shapira E, Blitzer MG, Miller JB, Africk DK, eds. Biochemical genetics: A laboratory manual. Oxford, UK: Oxford University Press, 1989:94–5.
	2	Lepage N, McDonald N, Dallaire L, et al. Age-specific distribution of plasma amino acid concentrations in a healthy pediatric population. Clin Chem 1997;43:2397–402.
Method(s)	1	Beckman amino acid analyzer. Beckman 6300 (Beckman Instruments Inc., Palo Alto, CA).
	2	Beckman amino acid analyzer. Beckman 7300 (Beckman Instruments Inc., Palo Alto, CA).
Comment(s)	1	Results are mean ±2 SDs.
	2	Results are 10–90th percentiles in healthy children.
	*See references for numbers.	

AMINO ACIDS (PLASMA)

Test	Age	n	Male and Female (nmol/mL)			
			Asparagine	Aspartic Acid	Carnosine	Citrulline
1	Premature (first 6 wk)	*	90–295	24–50	—	20–87
	0–1 mo		29–132	20–129	0–19	10–45
	1–24 mo		21–95	0–23	0	3–35
	2–18 y		23–112	1–24	0	1–46
2	6 mo	*	31–56	4–18	—	14–32
	2 y		29–56	3–8	—	17–35
	6 y		31–67	3–6	—	23–37
	16 y		37–81	2–5	—	23–39

Specimen Type(s)	1,2	Plasma
Reference(s)	1	Shapira E, Blitzer MG, Miller JB, Africk DK, eds. Biochemical genetics: A laboratory manual. Oxford, UK: Oxford University Press, 1989:94–5.
	2	Lepage N, McDonald N, Dallaire L, et al. Age-specific distribution of plasma amino acid concentrations in a healthy pediatric population. Clin Chem 1997;43:2397–402.
Method(s)	1	Beckman amino acid analyzer. Beckman 6300 (Beckman Instruments Inc., Palo Alto, CA).
	2	Beckman amino acid analyzer. Beckman 7300 (Beckman Instruments Inc., Palo Alto, CA).
Comment(s)	1	Results are mean ±2 SDs.
	2	Results are 10–90th percentiles in healthy children.
	*See references for numbers.	

AMINO ACIDS (PLASMA)

			\multicolumn{4}{c}{Male and Female (nmol/mL)}			
Test	Age	n	Cystathionine	Cystine	Ethanolamine	Glutamic Acid
1	Premature (first 6 wk)	*	5–10	15–70	—	107–276
	0–1 mo		0–3	17–98	0–115	62–620
	1–24 mo		0–5	16–84	0–4	10–133
	2–18 y		0–3	5–45	0–7	5–150
2	6 mo	*	—	21–53	—	31–113
	2 y		—	27–52	—	28–81
	6 y		—	33–54	—	13–65
	16 y		—	36–61	—	11–46
Specimen Type(s)	1,2	Plasma				
Reference(s)	1	Shapira E, Blitzer MG, Miller JB, Africk DK, eds. Biochemical genetics: A laboratory manual. Oxford, UK: Oxford University Press, 1989;94–5.				
	2	Lepage N, McDonald N, Dallaire L, et al. Age-specific distribution of plasma amino acid concentrations in a healthy pediatric population. Clin Chem 1997;43:2397–402.				
Method(s)	1	Beckman amino acid analyzer. Beckman 6300 (Beckman Instruments Inc., Palo Alto, CA).				
	2	Beckman amino acid analyzer. Beckman 7300 (Beckman Instruments Inc., Palo Alto, CA).				
Comment(s)	1	Results are mean ±2 SDs.				
	2	Results are 10–90th percentiles in healthy children.				
	*See references for numbers.					

AMINO ACIDS (PLASMA)

			Male and Female (nmol/mL)			
Test	Age	n	Glutamine	Glycine	Histidine	Homocystine
1	Premature (first 6 wk)	*	248–850	298–602	72–134	3–20
	0–1 mo		376–709	232–740	30–138	0
	1–24 mo		246–1182	81–436	41–101	0
	2–18 y		254–823	127–341	41–125	0–5
2	6 mo	*	474–737	138–276	61–91	—
	2 y		473–692	138–276	61–91	—
	6 y		493–724	144–282	63–93	—
	16 y		551–797	183–322	77–107	—

Specimen Type(s)	1,2	Plasma
Reference(s)	1	Shapira E, Blitzer MG, Miller JB, Africk DK, eds. Biochemical genetics: A laboratory manual. Oxford, UK: Oxford University Press, 1989;94–5.
	2	Lepage N, McDonald N, Dallaire L, et al. Age-specific distribution of plasma amino acid concentrations in a healthy pediatric population. Clin Chem 1997;43:2397–402.
Method(s)	1	Beckman amino acid analyzer. Beckman 6300 (Beckman Instruments Inc., Palo Alto, CA).
	2	Beckman amino acid analyzer. Beckman 7300 (Beckman Instruments Inc., Palo Alto, CA).
Comment(s)	1	Results are mean ±2 SDs.
	2	Results are 10–90th percentiles in healthy children.
	*See references for numbers.	

AMINO ACIDS (PLASMA)

			Male and Female (nmol/mL)			
Test	Age	n	Hydroxylysine	Hydroxyproline	Isoleucine	Leucine
1	Premature (first 6 wk)	*	0	0–80	23–85	151–220
	0–1 mo		0–7	0–91	26–91	48–160
	1–24 mo		0–7	0–63	31–86	47–155
	2–18 y		0–2	3–45	22–107	49–216
2	6 mo	*	—	—	39–76	77–153
	2 y		—	—	4–78	79–147
	6 y		—	—	40–69	86–136
	16 y		—	—	47–74	101–159

Specimen Type(s)	1,2	Plasma
Reference(s)	1	Shapira E, Blitzer MG, Miller JB, Africk DK, eds. Biochemical genetics: A laboratory manual. Oxford, UK: Oxford University Press, 1989;94–5.
	2	Lepage N, McDonald N, Dallaire L, et al. Age-specific distribution of plasma amino acid concentrations in a healthy pediatric population. Clin Chem 1997;43:2397–402.
Method(s)	1	Beckman amino acid analyzer. Beckman 6300 (Beckman Instruments Inc., Palo Alto, CA).
	2	Beckman amino acid analyzer. Beckman 7300 (Beckman Instruments Inc., Palo Alto, CA).
Comment(s)	1	Results are mean ±2 SDs.
	2	Results are 10–90th percentiles in healthy children.
	*See references for numbers.	

AMINO ACIDS (PLASMA)

					1-Methyl-	3-Methyl-
Test	Age	n	Lysine	Methionine	histidine	histidine
1	Premature (first 6 wk)	*	128–255	37–91	4–28	5–33
	0–1 mo		92–325	10–60	0–43	0–5
	1–24 mo		52–196	9–42	0–44	0–5
	2–18 y		48–284	7–47	0–42	0–5
2	6 mo	*	87–171	14–38	—	—
	2 y		88–172	13–22	—	—
	6 y		96–181	14–25	—	—
	16 y		157–242	20–34	—	—

Specimen Type(s)	1,2	Plasma
Reference(s)	1	Shapira E, Blitzer MG, Miller JB, Africk DK, eds. Biochemical genetics: A laboratory manual. Oxford, UK: Oxford University Press, 1989;94–5.
	2	Lepage N, McDonald N, Dallaire L, et al. Age-specific distribution of plasma amino acid concentrations in a healthy pediatric population. Clin Chem 1997;43:2397–402.
Method(s)	1	Beckman amino acid analyzer. Beckman 6300 (Beckman Instruments Inc., Palo Alto, CA).
	2	Beckman amino acid analyzer. Beckman 7300 (Beckman Instruments Inc., Palo Alto, CA).
Comment(s)	1	Results are mean ±2 SDs.
	2	Results are 10–90th percentiles in healthy children.
	*See references for numbers.	

AMINO ACIDS (PLASMA)

	Male and Female (nmol/mL)					
Test	Age	n	Ornithine	Phenylalanine	Phospho-ethanolamine	Phosphoserine
1	Premature (first 6 wk)	*	77–212	98–213	5–35	10–45
	0–1 mo		48–211	38–137	3–27	7–47
	1–24 mo		22–103	31–75	0–6	1–20
	2–18 y		10–163	26–91	0–69	1–30
2	6 mo	*	25–103	38–78	—	—
	2 y		24–60	39–65	—	—
	6 y		25–50	40–61	—	—
	16 y		37–62	47–74	—	—
Specimen Type(s)	1,2	Plasma				
Reference(s)	1	Shapira E, Blitzer MG, Miller JB, Africk DK, eds. Biochemical genetics: A laboratory manual. Oxford, UK: Oxford University Press, 1989;94–5.				
	2	Lepage N, McDonald N, Dallaire L, et al. Age-specific distribution of plasma amino acid concentrations in a healthy pediatric population. Clin Chem 1997;43:2397–402.				
Method(s)	1	Beckman amino acid analyzer. Beckman 6300 (Beckman Instruments Inc., Palo Alto, CA).				
	2	Beckman amino acid analyzer. Beckman 7300 (Beckman Instruments Inc., Palo Alto, CA).				
Comment(s)	1	Results are mean ±2 SDs.				
	2	Results are 10–90th percentiles in healthy children.				
	*See references for numbers.					

AMINO ACIDS (PLASMA)

Test	Age	n	Proline	Sarcosine	Serine	Taurine
			Male and Female (nmol/mL)			
1	Premature (first 6 wk)	*	92–310	0	127–248	151–411
	0–1 mo		110–417	0–625	99–395	46–492
	1–24 mo		52–298	0	71–186	15–143
	2–18 y		59–369	0–9	69–187	10–170
2	6 mo	*	93–265	—	98–160	39–111
	2 y		93–220	—	97–154	39–80
	6 y		93–201	—	96–155	41–69
	16 y		113–271	—	101–177	41–66

Specimen Type(s)	1,2	Plasma
Reference(s)	1	Shapira E, Blitzer MG, Miller JB, Africk DK, eds. Biochemical genetics: A laboratory manual. Oxford, UK: Oxford University Press, 1989;94–5.
	2	Lepage N, McDonald N, Dallaire L, et al. Age-specific distribution of plasma amino acid concentrations in a healthy pediatric population. Clin Chem 1997;43:2397–402.
Method(s)	1	Beckman amino acid analyzer. Beckman 6300 (Beckman Instruments Inc., Palo Alto, CA).
	2	Beckman amino acid analyzer. Beckman 7300 (Beckman Instruments Inc., Palo Alto, CA).
Comment(s)	1	Results are mean ±2 SDs.
	2	Results are 10–90th percentiles in healthy children.
	*See references for numbers.	

AMINO ACIDS (PLASMA)

			Male and Female (nmol/mL)			
Test	Age	n	Threonine	Tryptophan	Tyrosine	Valine
1	Premature (first 6 wk)	*	150–330	28–136	147–420	99–220
	0–1 mo		90–329	0–60	55–147	86–190
	1–24 mo		24–174	23–71	22–108	64–294
	2–18 y		35–226	0–79	24–115	74–321
2	6 mo	*	61–162	34–73	43–108	135–260
	2 y		61–115	35–73	40–77	147–255
	6 y		65–125	37–76	39–65	165–234
	16 y		104–188	54–93	46–87	178–275

Specimen Type(s)	1,2	Plasma
Reference(s)	1	Shapira E, Blitzer MG, Miller JB, Africk DK, eds. Biochemical genetics: A laboratory manual. Oxford, UK: Oxford University Press, 1989;94–5.
	2	Lepage N, McDonald N, Dallaire L, et al. Age-specific distribution of plasma amino acid concentrations in a healthy pediatric population. Clin Chem 1997;43:2397–402.
Method(s)	1	Beckman amino acid analyzer. Beckman 6300 (Beckman Instruments Inc., Palo Alto, CA).
	2	Beckman amino acid analyzer. Beckman 7300 (Beckman Instruments Inc., Palo Alto, CA).
Comment(s)	1	Results are mean ±2 SDs.
	2	Results are 10–90th percentiles in healthy children.
	*See references for numbers.	

AMINO ACIDS (URINE)

Male and Female (nmol/mg Creatinine)

Age	n	Alanine	β-Alanine	Anserine	α-Aminoadipic Acid
Premature (first 6 wk)	*	1320–4040	1020–3500	—	70–460
0–1 mo		982–3055	25–288	0–3	0–180
1–24 mo		767–6090	23–71	0–5	45–268
2–18 y		231–915	0–65	0	2–88

Specimen Type(s)	Urine
Reference(s)	Shapira E, Blitzer MG, Miller JB, Africk DK, eds. Biochemical genetics: a laboratory manual. Oxford, UK: Oxford University Press, 1989:96–7.
Method(s)	Beckman amino acid analyzer. Beckman 6300 (Beckman Instruments Inc., Palo Alto, CA).
Comment(s)	Results are mean ±2 SDs. *See references for numbers.

AMINO ACIDS (URINE)

Male and Female (nmol/mg Creatinine)

Age	n	α-Amino-n-butyric Acid	γ-Amino-butyric Acid	β-Amino-isobutyric Acid	Arginine
Premature (first 6 wk)	*	50–710	20–260	50–470	190–820
0–1 mo		8–65	0–15	421–3133	35–214
1–24 mo		30–136	0–105	802–4160	38–165
2–18 y		0–77	15–30	291–1482	31–109

Specimen Type(s)	Urine
Reference(s)	Shapira E, Blitzer MG, Miller JB, Africk DK, eds. Biochemical genetics: a laboratory manual. Oxford, UK: Oxford University Press, 1989:96–7.
Method(s)	Beckman amino acid analyzer. Beckman 6300 (Beckman Instruments Inc., Palo Alto, CA).
Comment(s)	Results are mean ±2 SDs. *See references for numbers.

AMINO ACIDS (URINE)

Male and Female (nmol/mg Creatinine)					
Age	n	Asparagine	Aspartic Acid	Carnosine	Citrulline
Premature (first 6 wk)	*	1350–5250	580–1520	260–370	240–1320
0–1 mo		185–1550	336–810	97–665	27–181
1–24 mo		252–1280	230–685	203–635	22–180
2–18 y		72–332	0–120	72–402	10–99

Specimen Type(s)	Urine
Reference(s)	Shapira E, Blitzer MG, Miller JB, Africk DK, eds. Biochemical genetics: a laboratory manual. Oxford, UK: Oxford University Press, 1989:96–7.
Method(s)	Beckman amino acid analyzer. Beckman 6300 (Beckman Instruments Inc., Palo Alto, CA).
Comment(s)	Results are mean ±2 SDs. *See references for numbers.

AMINO ACIDS (URINE)

Male and Female (nmol/mg Creatinine)					
Age	n	Cystathionine	Cystine	Ethanolamine	Glutamic Acid
Premature (first 6 wk)	*	260–1160	480–1690	—	380–3760
0–1 mo		16–147	212–668	840–3400	70–1058
1–24 mo		33–470	68–710	0–2230	54–590
2–18 y		0–26	25–125	0–530	0–176

Specimen Type(s)	Urine
Reference(s)	Shapira E, Blitzer MG, Miller JB, Africk DK, eds. Biochemical genetics: a laboratory manual. Oxford, UK: Oxford University Press, 1989:96–7.
Method(s)	Beckman amino acid analyzer. Beckman 6300 (Beckman Instruments Inc., Palo Alto, CA).
Comment(s)	Results are mean ±2 SDs. *See references for numbers.

AMINO ACIDS (URINE)

Male and Female (nmol/mg Creatinine)					
Age	n	Glutamine	Glycine	Histidine	Homocystine
Premature (first 6 wk)	*	520–1700	7840–23600	1240–7240	580–2230
0–1 mo		393–1042	5749–16423	908–2528	0–88
1–24 mo		670–1562	3023–11148	815–7090	6–67
2–18 y		369–1014	897–4500	644–2430	0–32

Specimen Type(s)	Urine
Reference(s)	Shapira E, Blitzer MG, Miller JB, Africk DK, eds. Biochemical genetics: a laboratory manual. Oxford, UK: Oxford University Press, 1989:96–7.
Method(s)	Beckman amino acid analyzer. Beckman 6300 (Beckman Instruments Inc., Palo Alto, CA).
Comment(s)	Results are mean ±2 SDs. *See references for numbers.

AMINO ACIDS (URINE)

Male and Female (nmol/mg Creatinine)					
Age	n	Hydroxylysine	Hydroxyproline	Isoleucine	Leucine
Premature (first 6 wk)	*	—	560–5640	250–640	190–790
0–1 mo		10–125	40–440	125–390	78–195
1–24 mo		0–97	0–4010	38–342	70–570
2–18 y		40–102	0–3300	10–126	30–500

Specimen Type(s)	Urine
Reference(s)	Shapira E, Blitzer MG, Miller JB, Africk DK, eds. Biochemical genetics: a laboratory manual. Oxford, UK: Oxford University Press, 1989:96–7.
Method(s)	Beckman amino acid analyzer. Beckman 6300 (Beckman Instruments Inc., Palo Alto, CA).
Comment(s)	Results are mean ±2 SDs. *See references for numbers.

AMINO ACIDS (URINE)

		\multicolumn{4}{c}{Male and Female (nmol/mg Creatinine)}			
Age	n	Lysine	Methionine	1-Methyl-histidine	3-Methyl-histidine
Premature (first 6 wk)	*	1860–15460	500–1230	170–880	420–1340
0–1 mo		270–1850	342–880	96–499	189–680
1–24 mo		189–850	174–1090	106–1275	147–391
2–18 y		153–634	16–114	170–1688	182–365
Specimen Type(s)	\multicolumn{5}{l}{Urine}				
Reference(s)	\multicolumn{5}{l}{Shapira E, Blitzer MG, Miller JB, Africk DK, eds. Biochemical genetics: a laboratory manual. Oxford, UK: Oxford University Press, 1989:96–7.}				
Method(s)	\multicolumn{5}{l}{Beckman amino acid analyzer. Beckman 6300 (Beckman Instruments Inc., Palo Alto, CA).}				
Comment(s)	\multicolumn{5}{l}{Results are mean ±2 SDs. *See references for numbers.}				

AMINO ACIDS (URINE)

Age	n	Ornithine	Phenylalanine	Phospho-ethanolamine	Phosphoserine
Premature (first 6 wk)	*	260–3350	920–2280	80–340	500–1690
0–1 mo		118–554	91–457	0–155	150–339
1–24 mo		55–364	175–1340	108–533	112–304
2–18 y		31–91	61–314	18–150	70–138
Specimen Type(s)	Urine				
Reference(s)	Shapira E, Blitzer MG, Miller JB, Africk DK, eds. Biochemical genetics: a laboratory manual. Oxford, UK: Oxford University Press, 1989:96–7.				
Method(s)	Beckman amino acid analyzer. Beckman 6300 (Beckman Instruments Inc., Palo Alto, CA).				
Comment(s)	Results are mean ±2 SDs. *See references for numbers.				

Male and Female (nmol/mg Creatinine)

AMINO ACIDS (URINE)

Male and Female (nmol/mg Creatinine)					
Age	n	Proline	Sarcosine	Serine	Taurine
Premature (first 6 wk)	*	1350–10460	0	1680–6000	5190–23620
0–1 mo		370–2323	0–56	1444–3661	1650–6220
1–24 mo		254–2195	30–358	845–3190	545–3790
2–18 y		0	0–26	362–1100	639–1866

Specimen Type(s)	Urine
Reference(s)	Shapira E, Blitzer MG, Miller JB, Africk DK, eds. Biochemical genetics: a laboratory manual. Oxford, UK: Oxford University Press, 1989:96–7.
Method(s)	Beckman amino acid analyzer. Beckman 6300 (Beckman Instruments Inc., Palo Alto, CA).
Comment(s)	Results are mean ±2 SDs. *See references for numbers.

AMINO ACIDS (URINE)

Male and Female (nmol/mg Creatinine)					
Age	n	Threonine	Tryptophan	Tyrosine	Valine
Premature (first 6 wk)	*	840–5700	0	1090–6780	180–890
0–1 mo		445–1122	0	220–1650	113–369
1–24 mo		252–1528	0–93	333–1550	99–316
2–18 y		121–389	0–108	122–517	58–143

Specimen Type(s)	Urine
Reference(s)	Shapira E, Blitzer MG, Miller JB, Africk DK, eds. Biochemical genetics: a laboratory manual. Oxford, UK: Oxford University Press, 1989:96–7.
Method(s)	Beckman amino acid analyzer. Beckman 6300 (Beckman Instruments Inc., Palo Alto, CA).
Comment(s)	Results are mean ±2 SDs. *See references for numbers.

AMMONIA

	Male and Female		
Test	Age	n	μmol/L
1	All ages	*	<33
2	Newborn	**	<50
	Child and adult		<35
3	<30 d	87	21–95
	1–12 mo	91	18–74
	1–14 y	94	17–68
	<14 y	182	22–66
	Men	89	21–71
	Women	93	19–63
4	All ages	***	<51

Specimen Type(s)	1,2	Plasma
	3	Whole blood, potassium EDTA as anticoagulant
	4	Plasma/serum
Reference(s)	1	Children's National Medical Center. Unpublished data.
	2	Reference values and SI unit information. The Hospital for Sick Children, Toronto, 1993.
	3	Diaz J, Tornel PL, Martinez P. Reference intervals for blood ammonia in healthy subjects, determined by microdiffusion. Clin Chem 1995;41:1048. (Letter)
	4	Ghoshal AK, Soldin SJ. Evaluation of the Dade Behring Dimension RxL: integrated chemistry system-pediatric reference ranges. Clin Chim Acta 2003;331:135–46.
Method(s)	1,2	Vitros 500 and 700 (Ortho-Clinical Diagnostics, Raritan, NJ).
	3	Ammonia Checker II (Menarini Diagnostics, Florence, Italy).
	4	Glutamate dehydrogenase (GLDH) catalyzes the condensation of ammonia and α-ketoglutarate (α-KG) with simultaneous oxidation of reduced nicotinamide adenine dinucleotide phosphate (NADPH). Siemens Dimension RxL Analyzer (Siemens Healthcare Diagnostics, Deerfield, IL).
Comment(s)	1	Values in premature infants may go as high as 50 μmol/L. Values are 2.5–97.5th percentiles.
		*Numbers not given.
	2	**Numbers not given.
	3	Values provided are 5–95th percentiles.
	4	***Reference ranges were obtained by comparing results from previously published data and using regression equations.

\multicolumn{4}{c}{**AMYLASE**}			
\multicolumn{4}{c}{Male and Female}			
Test	Age	n	U/L
1	1–30 d	76	0–6
	31–182 d	110	1–17
	183–365 d	54	6–44
	1–3 y	148	8–79
	4–9 y	96	16–91
	10–18 y	142	19–76
2	0–0.2 y	55	<30
	0.2–0.5 y	81	<50
	0.5–1.0 y	170	<80
3	1–19 y	470	30–100
4	1–30 d	*	<18
	31–182 d		<43
	183–365 d		<81
	1–3 y		<106
	4–9 y		<106
	10–18 y		<106

Specimen Type(s)	1,2	Plasma
	3	Serum
	4	Plasma/serum
Reference(s)	1	Soldin SJ, Hicks JM, Bailey J, et al. Pediatric reference ranges for amylase. Clin Chem 1995;41:S94. (Abstract)
	2	Soldin SJ, Rakotoarisoa FTS. Pediatric reference ranges for amylase and cholesterol on the Kodak Ektachem 500 in the first year of life. Clin Chem 1996;42:S308. (Abstract)
	3	Lockitch G, Halstead AC, Albersheim S, et al. Age- and sex-specific pediatric reference intervals for biochemistry analytes as measured with the Ektachem-700 analyzer. Clin Chem 1988;34:1622–5.
	4	Ghoshal AK, Soldin SJ. Evaluation of the Dade Behring Dimension RxL: integrated chemistry system-pediatric reference ranges. Clin Chim Acta 2003;331:135–46.
Method(s)	1	Hitachi 717 using Boehringer Mannheim reagents (Boehringer Mannheim Diagnostics, Indianapolis, IN).
	2,3	Amylopectin. Vitros 500 and 700 (Ortho-Clinical Diagnostics, Raritan, NJ).
	4	The amylase method on the Dimension® system uses a chromogenic substrate, 2-chloro-4-nitrophenol linked with maltotriose. The direct reaction of α-amylase with the substrate results in the formation of 2-chloro-4 nitrophenol, which is monitored spectrophotometrically. Siemens Dimension RxL Analyzer (Siemens Healthcare Diagnostics, Deerfield, IL).
Comment(s)	1	Study used hospitalized patients and a computerized approach to removing outliers. Values are 2.5–97.5th percentiles.
	2	Study used hospitalized patients and a computerized approach to removing outliers. Values are 97.5th percentiles.
	3	The study population was healthy children. Non-parametric methods were used to determine the 0.025 and 0.975 fractiles.
	4	*Reference ranges were obtained by comparing results from previously published data and using regression equations.

ANDROSTENEDIONE

Test	Age	Male n	Male ng/dL	Male nmol/L	Female n	Female ng/dL	Female nmol/L
1	1–5 mo	9	5–45	0.2–1.6	5	5–35	0.2–1.2
	6–11 mo	8	5–30	0.2–1.0	6	5–25	0.2–0.9
	1–5 y	31	5–45	0.2–1.6	18	5–40	0.2–1.4
	6–9 y	17	5–55	0.2–1.9	9	5–45	0.2–1.6
	10–11 y	7	10–30	0.3–1.0	11	25–80	0.9–2.8
	12–14 y	31	20–85	0.7–3.0	23	15–175	0.5–6.1
	15–17 y	16	35–100	1.2–3.5	10	55–200	1.9–7.0
	Adults	*	50–250	1.7–8.7	*	50–250	1.7–8.7
2	0–1 d	**	15–145	0.5–5.0	**	15–175	0.5–6.0
	1–7 d		20–110	0.7–3.8		25–95	0.9–3.3
	7–28 d		25–160	0.9–5.5		9–90	0.3–3.0
	1–12 mo		6–90	0.2–3.0		6–145	0.2–5.0
	1–<4 y		6–35	0.2–1.2		6–45	0.2–1.5
	4–<10 y						
	P_1		25–190	0.8–3.0		3–60	0.1–2.0
	P_2		15–120	0.5–4.0		30–145	1.0–5.0
	P_3		18–145	0.6–5.0		30–200	1.0–7.0
	P_4		15–220	0.5–7.5		18–260	0.6–9.0
	P_5		40–260	1.3–9.0		18–260	0.6–9.0

Test	Male and Female			
	Age	n	ng/dL	nmol/L
3	Premature infants	***		
	26–28 w, day 4		92–282	3.2–9.8
	31–35 w, day 4		80–446	2.8–15.5
	Full-term infants			
	1–7 d[a]		20–290[b]	0.7–10.1
	1–12 mo[c]		6–68	0.2–2.4
	Prepubertal children			
	1–10 y		8–50	0.3–1.7

Male			Female[d]		
Age	ng/dL	nmol/L	Age	ng/dL	nmol/L
Puberty			Puberty		
Tanner Stage[e]			Tanner Stage[e]		
1 (<9.8 y)	8–50	0.3–1.7	1 (<9.2 y)	8–50	0.3–1.7
2 (9.8–14.5 y)	31–65	1.1–2.3	2 (9.2–13.7 y)	42–100	1.5–3.5
3 (10.7–15.4 y)	50–100	1.7–3.5	3 (10.0–14.4 y)	80–190	2.8–6.6
4 (11.8–16.2 y)	48–140	1.7–4.9	4 (10.7–15.6 y)	77–225	2.7–7.8
5 (12.8–17.3 y)	65–210	2.3–7.3	5 (11.8–18.6 y)	80–240	2.8–8.4
n***			n***		

Specimen Type(s)	1,3	Serum
	2	Plasma/serum
Reference(s)	1	Pediatric endocrine testing. Nichols Institute, 1993:5.
	2	Soldin SJ, Rifai N, Hicks JM, eds. Biochemical basis of pediatric disease, 3rd ed. Washington, DC: AACC Press, 1998:233.
	3	Endocrinology expected values. Esoterix Endocrinology, Calabasas Hills, CA. ©2002 Esoterix, Inc., http://www.esoterix.com
		[a]Forest MG, Bertrand J. Sexual steroids in neonatal period. Abstracts of the 7th Congress of the International Study Group for Steroids. J Steroid Biochem 1975;6:xxiv.
		[e]Tanner JM, Whitehouse RH. Clinical longitudinal standards for height, weight, height velocity, and the stages of puberty. Arch Dis Childhood 1976;51:170.
Method(s)	1	Extraction, chromatography, radioimmunoassay.
	2	See reference.
	3	Not provided.
Comment(s)	1	*Numbers not provided.
		Results are 2.5–97.5th percentiles.
	2	**Numbers not provided.
		P_1–P_5 refers to pubertal stages. These ranges are for guidance only.
	3	***Numbers not provided.
		[b]Levels decrease rapidly to a range of 18–80 ng/dL after one week.
		[c]Androstenedione gradually decreases during the first six months to prepubertal levels.
		[d]Day of cycle not determined for prepubertal females.
		[e]As used herein, Tanner stages in males encompass development of both pubic hair and genitalia. In females, each stage encompasses both pubic hair and breast development. While this expands the chronological age range for each stage of pubertal development, it results in a better correlation with hormonal values.

APOLIPOPROTEIN A-1

Test	Age	Male			Female		
		n	g/L	mg/dL	n	g/L	mg/dL
1	5 y	*	1.04–1.67	104–167	*	1.08–1.60	108–160
	10 y		1.05–1.60	105–160		1.04–1.57	104–157
	15 y		1.02–1.57	102–157		1.04–1.56	104–156
	20 y		1.04–1.65	104–165		1.08–1.65	108–165
2	4–5 y	**	1.09–1.72	109–172	**	1.04–1.63	104–163
	6–11 y		1.11–1.77	111–177		1.10–1.66	110–166
	12–19 y		0.99–1.65	99–165		1.05–1.80	105–180

Test	Age	Male and Female		
		n	g/L	mg/dL
3	2–12 mo	100	0.79–1.87	79–187
	2–10 y	120	1.07–1.79	107–179

Test	Age	Male			Female		
		n	g/L	mg/dL	n	g/L	mg/dL
4	0–1 y	55	0.59–1.75	59–175	72	0.85–1.79	85–179

Age	Male and Female		
	n	g/L	mg/dL
1–5 y	144	0.83–1.74	83–174
6–10 y	131	1.03–1.77	103–177
11–14 y	154	0.99–1.80	99–180

Age	Male			Female		
	n	g/L	mg/dL	n	g/L	mg/dL
15–20 y	36	0.88–1.74	88–174	143	1.03–1.77	103–177

Test	Age	Male and Female		
		n	g/L	mg/dL
5	0–≤1 y	99	1.00–1.86	100–186
	>1–≤15 y	404	0.92–1.83	92–183

Age	Male			Female		
	n	g/L	mg/dL	n	g/L	mg/dL
>15–≤20 y	35	0.88–1.64	88–164	50	0.89–2.03	89–203

Specimen Type(s)	1	Plasma
	2,3	Serum
	4,5	Plasma/serum
Reference(s)	1	Kottke BA, Moll PP, Michels VV, et al. Levels of lipids, lipoproteins, and apolipoproteins in a defined population. Mayo Clin Proc 1991;66:1198–208.
	2	Bachorik PS, Lovejoy KL, Carroll MD, et al. Apolipoprotein B and A1 distributions in the United States, 1988–1991: results of the National Health and Nutrition Examination Survey III (NHANES III). Clin Chem 1997;43:2364–78.
	3	Baroni S, Scribano D, Valentini P, et al. Serum apolipoprotein A1, B, CII, CIII, E, and lipoprotein (a) levels in children. Clin Biochem 1996;29:603–5.
	4	Chan MK, Seiden-Long I, Aytekin M, et al. Canadian Laboratory Initiative on Reference Interval Database (CALIPER): pediatric reference intervals for an integrated clinical chemistry and immunoassay analyzer, Abbott ARCHITECT ci8200. Clin Biochem 2009;42:885–91.
	5	Kulasingam V, Jung BP, Blasutig IM, et al. Pediatric reference intervals for 28 chemistries and immunoassays on the Roche cobas® 6000 analyzer—A CALIPER pilot study. Clin Biochem 2010;43;1045–50.
Method(s)	1	Radioimmunoassay. See reference.
	2	Radial immunodiffusion and rate immunonephelometry. See reference above.
	3	Behring nephelometry using Behring reagents (Behring Diagnostics, Westwood, MA).
	4	Abbott Architect ci8200 (Abbott Diagnostics, Abbott Park, IL).
	5	Roche cobas® 6000 analyzer (Roche Diagnostics Limited, West Sussex, UK).
Comment(s)	1	Results are 5–95th percentiles read off graphs in above reference. *See references for numbers.
	2	Healthy children were studied. The results are the 5–95th percentiles. **The numbers studied are not provided.
	3	Results are the 2.5–97.5th percentiles for healthy children.
	4	1459 serum/plasma specimens from children attending select outpatient clinics were collected from the five age groups noted above. Values are 2.5–97.5th percentiles.
	5	Approximately 600 outpatient samples from a pediatric population deemed to be metabolically stable were subdivided into five age classes ranging from 0 to 20 years of age and further partitioned by gender. Values are 2.5–97.5th percentiles.

APOLIPOPROTEIN (a)

	Male and Female		
Test	Age	n	U/L
1	2 d	29	0–59
	3 d	151	0–57
	4 d	459	0–69
	5 d	341	0–66
	6 d	30	0–78
	7 d	22	0–71
Test	Age	n	mg/dL
2	2–12 mo	100	0–114
	2–10 y	120	0–178

Specimen Type(s)	1	Dried blood spot
	2	Serum
Reference(s)	1	Wang XL, Wilcken DEL, Dudman NPB. Neonatal apo A-1, apo-B and apo(a) levels in dried blood spots in an Australian population. Ped Res 1990;28:496–501.
	2	Baroni S, Scribano D, Valentini P, et al. Serum apolipoprotein A1, B, CII, CIII, E and lipoprotein (a) levels in children. Clin Biochem 1996;29:603–5.
Method(s)	1	Modification of RIA serum assay for apo(a) (Pharmacia Diagnostics AB, Uppsala, Sweden).
	2	EIA, Macra Lpa test kit (Strategic Diagnostics, Newark, DE).
Comment(s)	1	Healthy neonates were studied. Results are mean ±2 SDs.
	2	Results are the 2.5–97.5th percentiles for healthy children.

APOLIPOPROTEIN B

Test	Age	Male			Female		
		n	g/L	mg/dL	n	g/L	mg/dL
1	5 y	*	0.51–0.80	51–80	*	0.54–0.87	54–87
	10 y		0.50–0.88	50–88		0.54–0.87	54–87
	15 y		0.48–0.85	48–85		0.54–0.87	54–87
	20 y		0.48–0.90	48–90		0.54–0.85	54–85
2	4–5 y	**	0.58–1.03	58–103	**	0.58–1.04	58–104
	6–11 y		0.56–1.05	56–105		0.57–1.13	57–113
	12–19 y		0.55–1.10	55–110		0.53–1.19	53–119

Test	Age	Male and Female		
		n	g/L	mg/dL
3	2–12 mo	100	0.41–1.05	41–105
	2–10 y	120	0.44–1.12	44–112
4	0–12 mo	124	0.22–1.16	22–116
	1–5 y	152	0.48–1.09	48–109
	6–10 y	131	0.45–1.04	45–104
	11–14 y	154	0.43–1.13	43–113
	15–20 y	179	0.00–1.20	0–120
5	0–<1 y	90	<0.03***–1.09	<3***–109
	≥1–≤20 y	470	0.38–1.05	38–105

Specimen Type(s)	1	Plasma
	2,3	Serum
	4,5	Plasma/serum
Reference(s)	1	Kottke BA, Moll PP, Michels VV, et al. Levels of lipids, lipoproteins, and apolipoproteins in a defined population. Mayo Clin Proc 1991;66:1198–208.
	2	Bachorik PS, Lovejoy KL, Carroll MD, et al. Apolipoprotein B and A1 distributions in the United States, 1988–1991: results of the National Health and Nutrition Examination Survey III (NHANES III). Clin Chem 1997;43:2364–78.
	3	Baroni S, Scribano D, Valentini P, et al. Serum apolipoprotein A1, B, CII, CIII, E, and lipoprotein (a) levels in children. Clin Biochem 1996;29:603–5.
	4	Chan MK, Seiden-Long I, Aytekin M, et al. Canadian Laboratory Initiative on Reference Interval Database (CALIPER): pediatric reference intervals for an integrated clinical chemistry and immunoassay analyzer, Abbott ARCHITECT ci8200. Clin Biochem 2009;42:885–91.
	5	Kulasingam V, Jung BP, Blasutig IM, et al. Pediatric reference intervals for 28 chemistries and immunoassays on the Roche cobas® 6000 analyzer—A CALIPER pilot study. Clin Biochem 2010;43;1045–50.
Method(s)	1	Enzyme linked immunosorbent assay. See reference.
	2	Radial immunodiffusion and rate nephelometry. See reference above.
	3	Behring nephelometer using Behring reagents (Behring Diagnostics, Westwood, MA).
	4	Abbott Architect ci8200 (Abbott Diagnostics, Abbott Park, IL).
	5	Roche cobas® 6000 analyzer (Roche Diagnostics Limited, West Sussex, UK).
Comment(s)	1	Results are 5–95th percentiles read off graphs in above reference. *See references for numbers.
	2	Healthy children were studied. The results are the 5–95th percentiles. **The numbers studied are not provided.
	3	Results are the 2.5–97.5th percentiles for healthy children.
	4	1459 serum/plasma specimens from children attending select outpatient clinics were collected from the five age groups noted above. Values are 2.5–97.5th percentiles.
	5	***Lower reference interval at the limit of detection of the assay. Approximately 600 outpatient samples from a pediatric population deemed to be metabolically stable were subdivided into five age classes ranging from 0 to 20 years of age and further partitioned by gender. Values are 2.5–97.5th percentiles.

APOLIPOPROTEIN CII

	Male and Female	
Age	n	mg/L
2–12 mo	100	15–79
2–10 y	120	9–73

Specimen Type(s)	Serum
Reference(s)	Baroni S, Scribano D, Valentini P, et al. Serum apolipoprotein A1, B, CII, CIII, E, and lipoprotein (a) levels in children. Clin Biochem 1996;29:603–5.
Method(s)	Turbidimetric immunoassay (Turbilinear "Eiken" Poli, Milan, Italy).
Comment(s)	Results are the 2.5–97.5th percentiles for healthy children.

APOLIPOPROTEIN CIII

	Male and Female	
Age	n	mg/L
2–12 mo	100	18–134
2–10 y	120	25–113

Specimen Type(s)	Serum
Reference(s)	Baroni S, Scribano D, Valentini P, et al. Serum apolipoprotein A1, B, CII, CIII, E, and lipoprotein (a) levels in children. Clin Biochem 1996;29:603–5.
Method(s)	Turbidimetric immunoassay (Turbilinear "Eiken" Poli, Milan, Italy).
Comment(s)	Results are the 2.5–97.5th percentiles for healthy children.

APOLIPOPROTEIN E

	Male and Female	
Age	n	mg/L
2–12 mo	100	23–59
2–10 y	120	19–59

Specimen Type(s)	Serum
Reference(s)	Baroni S, Scribano D, Valentini P, et al. Serum apolipoprotein A1, B, CII, CIII, E, and lipoprotein (a) levels in children. Clin Biochem 1996;29:603–5.
Method(s)	Turbidimetric immunoassay (Turbilinear "Eiken" Poli, Milan, Italy).
Comment(s)	Results are the 2.5–97.5th percentiles for healthy children.

ASPARTATE AMINOTRANSFERASE (AST, SGOT)

Test	Age	Male n	Male U/L	Female n	Female U/L
1	1–7 d	69	30–100	52	24–95
	8–30 d	148	20–70	84	24–72
	1–3 mo	160	22–63	131	20–64
	4–6 mo	133	13–65	83	20–63
	7–12 mo	131	25–55	142	22–63
2	1–3 y	50*	20–60	50*	20–60
	5–6 y	40*	15–50	40*	15–50
	7–9 y	80*	15–40	80*	15–40
	10–11 y	27	10–60	34	10–40
	12–13 y	31	15–40	49	10–30
	14–15 y	26	15–40	52	10–30
	16–19 y	40	15–45	61	5–30
3	1–30 d	74	<51	57	<49
	31–365 d	83	<65	71	<79
	1–3 y	134	<56	108	<69
	4–6 y	85	<48	84	<59
	7–9 y	122	<42	96	<41
	10–12 y	104	<38	62	<37
	13–15 y	88	<39	86	<32
	16–18 y	62	<39	78	<30
4	1–7 d	**	26–98	**	20–93
	8–30 d		16–67		20–69
	1–3 mo		16–60		16–61
	4–6 mo		16–62		16–60
	7–12 mo		16–52		16–60
	1–3 y		16–57		16–57
	5–6 y		10–47		10–47
	7–9 y		10–36		5–36
	12–15 y		10–36		5–26
	16–19 y		10–41		0–26

Specimen Type(s)	1,3,4	Plasma/serum
	2	Plasma
Reference(s)	1	Soldin SJ, Savwoir TV, Guo Y. Pediatric reference ranges for alkaline phosphatase, aspartate aminotransferase, and alanine aminotransferase in children less than 1 year old on the Vitros 500. Clin Chem 1997;43:S199. (Abstract)
	2	Lockitch G, Halstead AC, Albersheim S, et al. Age- and sex-specific pediatric reference intervals for biochemistry analytes as measured with the Ektachem-700 analyzer. Clin Chem 1988;34:1622–5.
	3	Soldin SJ, Hicks JM, Bailey J, et al. Pediatric reference ranges for AST. Clin Chem 1995;44:S94. (Abstract)
	4	Ghoshal AK, Soldin SJ. Evaluation of the Dade Behring Dimension RxL: integrated chemistry system-pediatric reference ranges. Clin Chim Acta 2003;331:135–46.
Method(s)	1,2	Vitros 500 (1) and 700 (2) (Ortho-Clinical Diagnostics, Raritan, NJ).
	3	Measured on the Hitachi 747 using Boehringer Mannheim reagents (Boehringer Mannheim Diagnostics, Indianapolis, IN).
	4	Aspartate aminotransferase catalyzes the transamination from L-aspartate to α-ketoglutarate, forming L-glutamate and oxalacetate. The oxalacetate formed is reduced to malate by malate dehydrogenase (MDH) with simultaneous oxidation of reduced nicotinamide adenine dinucleotide (NADH). Siemens Dimension RxL Analyzer (Siemens Healthcare Diagnostics, Deerfield, IL).
Comment(s)	1,3	Study used hospitalized patients and a computerized approach adapted from the Hoffmann technique to obtain the 2.5–97.5th percentiles (Test 1) and 97.5th percentile (Test 3).
	2	The study population was healthy children. Non-parametric methods were used to determine the 0.025 and 0.975 fractiles.
		*No significant differences were found for males and females. These ranges were therefore derived from combined data.
	4	**Reference ranges were obtained by comparing results from previously published data and using regression equations.

BASE EXCESS

		Male and Female	
Test	Age	n	mmol/L
1	Newborn	*	−10 to −2
	Infant		−7 to −1
	Child		−4 to +2
	Thereafter		−3 to +3
2	Premature neonates	248	−6 to +7

Specimen Type(s)	1	Whole blood
	2	Capillary blood
Reference(s)	1	Behrman RE, ed. Nelson textbook of pediatrics, 14th ed. Philadelphia, PA: WB Saunders Company, 1992:1818.
	2	Soldin SJ. Children's National Medical Center. Unpublished data.
Method(s)	1	Not described.
	2	I-Stat (Abbott Laboratories, Abbott Park, IL).
Comment(s)	1	*Numbers not given.
	2	Study used hospitalized patients and a computerized approach adapted from the Hoffmann technique.

BICARBONATE (HCO_3^-)

	Male and Female	
Age	n	mmol/L
Premature neonates	248	22–31

Specimen Type(s)	Capillary blood
Reference(s)	Soldin SJ. Children's National Medical Center. Unpublished data.
Method(s)	I-Stat (Abbott Laboratories, Abbott Park, IL).
Comment(s)	Study used hospitalized patients and a computerized approach adapted from the Hoffmann technique.

BILE ACIDS (TOTAL)
(TOTAL 3-α-HYDROXY BILE ACIDS)

	Male and Female	
Age	n	μmol/L
1 mo	42	11–69
3 mo	35	1–41
6 mo	50	1–30
9 mo	36	1–28
12 mo	37	1–37
Adult	7	1–7

Specimen Type(s)	Serum
Reference(s)	McGraw C, Ellinor YM, Heubi JE. Reference ranges for total serum bile acids in infants. Clin Chem 1996;42:S307. (Abstract)
Method(s)	Nonradioactive enzymatic spectrophotometric method. "Enzabile," adapted for use on the Hitachi 705 analyzer (Boehringer Mannheim Diagnostics, Indianapolis, IN).
Comment(s)	Results are 2.5–97.5th percentiles. Population studied was a nonfasting, healthy population.

BILIRUBIN (CONJUGATED)

		Male and Female		
Test	Age	n	μmol/L	mg/dL
1	Neonates	*	<10	<0.6
	>Neonates		<2	<0.1
2	Preterm infants (1–6 d)	30	<10	<0.6
3	Neonates	**	<7	<0.4

Specimen Type(s)	1,2	Serum
	3	Plasma/serum
Reference(s)	1	Reference values and SI information. The Hospital for Sick Children, Toronto, Canada, 1993:359.
	2	Lockitch G, Halstead AC, Albersheim S, et al. Age- and sex-specific pediatric reference intervals for biochemistry analytes as measured with the Ektachem-700 analyzer. Clin Chem 1988;34:1622–5.
	3	Ghoshal AK, Soldin SJ. Evaluation of the Dade Behring Dimension RxL: integrated chemistry system-pediatric reference ranges. Clin Chim Acta 2003;331:135–46.
Method(s)	1,2	Vitros 700 (Ortho-Clinical Diagnostics, Raritan, NJ).
	3	Diazotized sulfanilic acid is formed by combining sodium nitrite and sulfanilic acid at low pH. Siemens Dimension RxL Analyzer (Siemens Healthcare Diagnostics, Deerfield, IL).
Comment(s)	1	*Numbers not given.
	2	Results are 97.5th percentile.
	3	**Reference ranges were obtained by comparing results from previously published data and using regression equations.

BILIRUBIN (TOTAL)

Test	Age	n	μmol/L	mg/dL
	Male and Female			
1	Birth–1 d	*	<100	<5.8
	1–2 d		<140	<8.2
	3–5 d		<200	<11.7
	1 mo–adult		<17	<1.0
2	Bottle fed infants	2416**	<212	<12.4
	Bottle fed infants	2416**	<253	<14.8
3	0–1 d	***	<87	<5.1
	1–2 d		<123	<7.2
	3–5 d		<176	<10.3
	1 mo–adult		<14	<0.8

Test	Age	Male			Female		
		n	μmol/L	mg/dL	n	μmol/L	mg/dL
4	1–12 mo	39	1.7–32.8	0.1–1.9	60	1.7–17.3	0.1–1.0

Age	n	μmol/L	mg/dL
Male and Female			
1–5 y	154	1.8–15.6	0.1–0.9
6–10 y	135	2.4–18.4	0.1–1.1
11–14 y	154	3.0–18.5	0.2–1.1

Age	Male			Female		
	n	μmol/L	mg/dL	n	μmol/L	mg/dL
15–20 y	35	4.1–34.4	0.2–2.0	148	3.3–23.4	0.2–1.4

Specimen Type(s)	1,2	Serum
	3,4	Plasma/serum
Reference(s)	1	Reference values and SI information. The Hospital for Sick Children, Toronto, Canada, 1993:359.
	2	Maisels MJ, Gifford K. Normal serum bilirubin levels in the newborn and the effect of breast-feeding. Pediatrics 1986;78:837–43.
	3	Ghoshal AK, Soldin SJ. Evaluation of the Dade Behring Dimension RxL: integrated chemistry system-pediatric reference ranges. Clin Chim Acta 2003;331:135–46.
	4	Chan MK, Seiden-Long I, Aytekin M, et al. Canadian Laboratory Initiative on Reference Interval Database (CALIPER): pediatric reference intervals for an integrated clinical chemistry and immunoassay analyzer, Abbott ARCHITECT ci8200. Clin Biochem 2009;42:885–91.
Method(s)	1	Vitros 700 (Ortho-Clinical Diagnostics, Raritan, NJ).
	2	Modified diazo method on ACA III (DuPont Co., Clinical Systems Division, Wilmington, DE).
	3	Diazotized sulfanilic acid is formed by combining sodium nitrite and sulfanilic acid at low pH. Bilirubin in the sample, including the delta form (1), is solubilized by dilution in a mixture of caffeine/benzoate/acetate/EDTA. Siemens Dimension RxL Analyzer (Siemens Healthcare Diagnostics Deerfield, IL). 1. Doumas BT, Wu T-W, Jendrzejczak B. Delta bilirubin: absorption spectra, molar absorptivity, and reactivity in the diazo reaction. Clin Chem 1987;33:769–74.
	4	Abbott Architect ci8200 (Abbott Diagnostics, Abbott Park, IL).
Comment(s)	1	Should decrease to adult values by day 10 (breast-fed infants may take longer). Values in premature infants may be higher than in term infants and may reach peak concentrations at later times. *Numbers not provided.
	2	**2416 consecutive infants admitted to the well-baby nursery studied. This number includes both formula- and breast-fed infants. Results are 97th percentile. Breast feeding was significantly associated with hyperbilirubinemia.
	3	***Reference ranges were obtained by comparing results from previously published data and using regression equations.
	4	1459 serum/plasma specimens from children attending select outpatient clinics were collected from the five age groups noted above. Values are 2.5–97.5th percentiles.

BRAIN NATRIURETIC PEPTIDE

		Male and Female	
Test	Age	n	ng/L (pg/mL)
1	0–<31 d	50	1585
	31–<90 d	38	1259
	3–<6 mo	26	759
	6 mo–<1 y	55	263

	Male		Female	
Age	n	ng/L (pg/mL)	n	ng/L (pg/mL)
1–<3 y	60	173	51	158
3–<10 y	89	132	72	120
10–<15 y	91	120	51	115
15–<18 y	63	100	66	107
18–21 y	50	110	46	87

		Male and Female	
Test	Age	n	ng/L (pg/mL)
2	0–12 mo	88	8–24
	1–5 y	143	8–30
	6–10 y	125	0.0–15
	11–14 y	169	7–21
	15–20 y	134	8–20

Specimen Type(s)	1	EDTA whole blood
	2	Plasma/serum
Reference(s)	1	Soldin SJ, Soldin OP, Boyajian AJ, Taskier MS. Pediatric brain natriuretic peptide and N-terminal pro-brain natriuretic peptide reference intervals. Clin Chim Acta 2006;366:304–8.
	2	Chan MK, Seiden-Long I, Aytekin M, et al. Canadian Laboratory Initiative on Reference Interval Database (CALIPER): pediatric reference intervals for an integrated clinical chemistry and immunoassay analyzer, Abbott ARCHITECT ci8200. Clin Biochem 2009;42:885–91.
Method(s)	1	Triage Biosite point-of-care method (Biosite Inc., San Diego, CA).
	2	Abbott Architect ci8200 (Abbott Diagnostics, Abbott Park, IL).
Comment(s)	1	Study used hospitalized patients and a computerized approach adapted from the Hoffmann technique. Values are 97.5th percentiles.
	2	1459 serum/plasma specimens from children attending select outpatient clinics were collected from the five age groups noted above. Values are 2.5–97.5th percentiles (rounded to the nearest whole number).

(N-TERMINAL PRO)-BRAIN NATRIURETIC PEPTIDE

Test	Age	Male		Female	
		n	ng/L (pg/mL)	n	ng/L (pg/mL)
1	0–<31 d	46	28184	53	35481
	31–<90 d	49	19953	45	15135
	3–<6 mo	49	15849	23	14125
	6 mo–<1 y	40	11220	50	10000
	1–<3 y	105	5012	59	2512
	3–<10 y	108	1259	90	1324
	10–<15 y	112	1585	87	1413
	15–<18 y	76	1584	103	1318
	18–21 y	61	1600	51	1400
2	0–<1 y	34	27–265	32	22–591

	Male and Female	
Age	n	ng/L (pg/mL)
≥1–≤5 y	120	12–308
≥5–≤15 y	194	8–178

	Male		Female	
Age	n	ng/L (pg/mL)	n	ng/L (pg/mL)
<15–≤20 y	20	<5*–74	33	7–137

Specimen Type(s)	1	Plasma
	2	Plasma/serum
Reference(s)	1	Soldin SJ, Soldin OP, Boyajian AJ, Taskier MS. Pediatric brain natriuretic peptide and N-terminal pro-brain natriuretic peptide reference intervals. Clin Chim Acta 2006;366:304–8.
	2	Kulasingam V, Jung BP, Blasutig IM, et al. Pediatric reference intervals for 28 chemistries and immunoassays on the Roche cobas® 6000 analyzer—A CALIPER pilot study. Clin Biochem 2010;43;1045–50.
Method(s)	1	Siemens Dimension RxL Analyzer (Siemens Healthcare Diagnostics, Deerfield, IL).
	2	Roche cobas® 6000 analyzer (Roche Diagnostics Limited, West Sussex,UK).
Comment(s)	1	Study used hospitalized patients and a computerized approach adapted from the Hoffmann technique. Values are 97.5th percentiles.
	2	*Lower reference interval at the limit of detection of the assay.
		Approximately 600 outpatient samples from a pediatric population deemed to be metabolically stable were subdivided into five age classes ranging from 0 to 20 years of age and further partitioned by gender. Values are 2.5–97.5th percentiles, rounded to the nearest whole number.

C-PEPTIDE

Test	Age	n	ng/mL	pmol/L
	Male and Female			
1	0–<1 y	73	0.6–7.8	199.8–2597.4
	1–<5 y	43	0.6–7.4	199.8–2464.2
	5–<13 y	75	0.8–8.5	266.4–2830.5
	13–<18y	76	1.3–7.9	432.9–2630.7
2	0–10 y	295	0.5–5.5	164.1–1841.1
	>10–≤15 y	147	0.7–11.2	230.6–3728.5

	Age	n	ng/mL	pmol/L	n	ng/mL	pmol/L
		Male			**Female**		
	>15–≤20 y	33	0.9–7.2	307.8–2402.7	47	0.9–9.4	284.6–3136.9

Specimen Type(s)	1,2	Plasma/serum
Reference(s)	1	Soldin OP, Dahlin JR, Gresham EG, et al. IMMULITE 2000 age- and sex-specific reference intervals for alpha fetoprotein, homocysteine, insulin, insulin-like growth factor-1, insulin-like growth factor binding protein-3, C-peptide, immunoglobulin E and intact parathyroid hormone. Clin Biochem 2008;41:937–42.
	2	Kulasingam V, Jung BP, Blasutig IM, et al. Pediatric reference intervals for 28 chemistries and immunoassays on the Roche cobas® 6000 analyzer—A CALIPER pilot study. Clin Biochem 2010;43;1045–50.
Method(s)	1	IMMULITE® 2000 immunoassay system. (Siemens Healthcare Diagnostics, Deerfield, IL).
	2	Roche cobas® 6000 analyzer (Roche Diagnostics Limited, West Sussex, UK).
Comment(s)	1	Study was conducted at both Children's National Medical Center and Georgetown University, Washington, DC. Results were obtained from the Children's National Medical Center laboratory information system over the period January 5, 2001, to March 8, 2007. Patient results were accessed and used to establish reference intervals. All patient identifiers were removed except age and sex. Data was analyzed using the Hoffmann approach, and was computer adapted. Values are 2.5–97.5th percentiles.
	2	Approximately 600 outpatient samples from a pediatric population deemed to be metabolically stable were subdivided into five age classes ranging from 0 to 20 years of age and further partitioned by gender. Values are 2.5–97.5th percentiles.

C1 ESTERASE INHIBITOR

Age	Male		Female	
	n	mg/L	n	mg/L
0–30 d	27	75–170	19	80–150
31–182 d	39	105–279	49	105–229
6 mo–<4 y	106	145–268	103	159–280
4–<7 y	75	140–252	48	120–269
7–<10 y	54	135–270	31	136–190
10–<13 y	60	118–209	61	142–211
13–<16 y	68	130–229	56	127–201
16–18 y	35	130–216	47	140–200

Specimen Type(s)	Plasma/serum
Reference(s)	Soldin SJ, Hicks JM, Bailey J, et al. Pediatric reference ranges for estradiol and C1 esterase inhibitor. Clin Chem 1998;44:A17. (Abstract)
Method(s)	DiaSorin kit (DiaSorin Corp., Stillwater, MN).
Comment(s)	The study used plasma/serum from hospitalized patients and employed Chauvenet's criteria for removing outliers and a computerized approach adapted from the Hoffmann technique to obtain the 2.5–97.5th percentiles.

CALCIUM

Test	Age	Male			Female		
		n	mg/dL	mmol/L	n	mg/dL	mmol/L
1	0–7 d	293	7.3–11.4	1.83–2.85	259	7.5–11.3	1.88–2.83
	8–30 d	434	8.6–11.7	2.15–2.93	264	8.4–11.9	2.10–2.98
	31–90 d	371	8.5–11.3	2.13–2.83	265	8.0–11.1	2.00–2.78
	91–180 d	186	8.3–11.4	2.08–2.85	222	7.7–11.5	1.93–2.88
	181–365 d	429	7.7–11.0	1.93–2.75	362	7.8–11.1	1.95–2.7
2	0–5 d (<2.5 kg)	50*	7.9–10.7	1.96–2.66	50*	7.9–10.7	1.96–2.66
	1–3 y	50*	8.7–9.8	2.17–2.44	50*	8.7–9.8	2.17–2.44
	4–6 y	38*	8.8–10.1	2.19–2.51	38*	8.8–10.1	2.19–2.51
	7–9 y	72*	8.8–10.1	2.19–2.51	72*	8.8–10.1	2.19–2.51
	10–11 y	62*	8.9–10.1	2.22–2.51	62*	8.9–10.1	2.22–2.51
	12–13 y	73*	8.8–10.6	2.19–2.64	73*	8.8–10.6	2.19–2.64
	14–15 y	91*	9.2–10.7	2.29–2.66	91*	9.2–10.7	2.29–2.66
	16–19 y	107*	8.9–10.7	2.22–2.66	107*	8.9–10.7	2.22–2.66
3	1–30 d	62	8.5–10.6	2.12–2.64	66	8.4–10.6	2.10–2.64
	31–365 d	83	8.7–10.5	2.17–2.62	66	8.9–10.5	2.22–2.62
	1–3 y	126	8.8–10.6	2.19–2.64	119	8.5–10.4	2.12–2.59
	4–6 y	112	8.8–10.6	2.19–2.64	106	8.5–10.6	2.12–2.64
	7–9 y	117	8.7–10.3	2.17–2.57	107	8.5–10.3	2.12–2.57
	10–12 y	135	8.7–10.2	2.17–2.54	115	8.6–10.2	2.15–2.64
	13–15 y	109	8.5–10.2	2.12–2.54	110	8.4–10.0	2.10–2.50
	16–18 y	95	8.4–10.3	2.10–2.57	122	8.6–9.8	2.15–2.45
4	0–12 mo	57	9.2–11.2	2.3–2.8	73	10.0–11.2	2.5–2.8

	Male and Female			
	Age	n	mg/dL	mmol/L
	1–5 y	152	9.2–10.4	2.3–2.6
	6–10 y	134	9.2–10.4	2.3–2.6
	11–14 y	154	8.8–10.4	2.2–2.6
	15–20 y	182	8.8–10.4	2.2–2.6

Test	Age	n	Male and Female			
			mg/dL		mmol/L	
5	0–<1 y	100	9.1–12.0		2.27–3.00	
	Age		Male		Female	
		n	mg/dL	mmol/L	n	mg/dL	mmol/L
	≥1–≤5 y	70	8.8–11.1	2.20–2.77	66	8.8–10.6	2.19–2.65
			Male and Female			
	Age	n	mg/dL		mmol/L	
	>5–≤10 y	106	8.6–10.4		2.16–2.61	
	Age		Male		Female	
		n	mg/dL	mmol/L	n	mg/dL	mmol/L
	>10–≤15 y	69	8.1–10.6	2.03–2.65	93	8.7–10.1	2.18–2.53
			Male and Female			
	Age	n	mg/dL		mmol/L	
	>15–≤20 y	85	8.6–10.5		2.16–2.62	

Test	Age	n	Male		Female	
			mg/dL	mmol/L	mg/dL	mmol/L
6	0–7 d	**	7.6–11.3	1.90–2.82	7.8–11.2	1.95–2.79
	8–30 d		8.8–11.6	2.20–2.89	8.6–11.8	2.15–2.94
	31–90 d		8.7–11.2	2.17–2.79	8.2–11.0	2.05–2.74
	91–180 d		8.5–11.3	2.12–2.82	8.0–11.4	2.00–2.84
	181–365 d		8.0–10.9	2.00–2.72	8.1–11.0	2.02–2.74
	1–3 y		8.9–9.9	2.22–2.47	8.9–9.9	2.22–2.47
	4–6 y		9.0–10.1	2.25–2.52	9.0–10.1	2.22–2.47
	7–9 y		9.0–10.1	2.25–2.52	9.0–10.1	2.22–2.47
	10–11 y		9.0–10.1	2.25–2.52	9.0–10.1	2.22–2.47
	12–13 y		9.0–10.6	2.25–2.64	9.0–10.6	2.22–2.64
	14–15 y		9.3–10.7	2.32–2.67	9.3–10.7	2.32–2.67
	16–19 y		9.0–10.7	2.25–2.67	9.0–10.7	2.25–2.67

Specimen Type(s)	1,3–6	Plasma/serum
	2	Serum
Reference(s)	1	Soldin SJ, Morse AS. Pediatric reference ranges for calcium and triglycerides in children <1 year old using the Vitros 500 analyzer. Clin Chem 1998;44:A16. (Abstract)
	2	Lockitch G, Halstead AC, Albersheim S, reference intervals for biochemistry analytes as measured with the Ektachem-700 analyzer. Clin Chem 1988;34:1622–5.
	3	Soldin SJ, Hicks JM, Bailey J, et al. Pediatric reference ranges for calcium on the Hitachi 747 Analyzer. Clin Chem 1997;43:S198. (Abstract)
	4	Chan MK, Seiden-Long I, Aytekin M, et al. Canadian Laboratory Initiative on Reference Interval Database (CALIPER): pediatric reference intervals for an integrated clinical chemistry and immunoassay analyzer, Abbott ARCHITECT ci8200. Clin Biochem 2009;42:885–91.
	5	Kulasingam V, Jung BP, Blasutig IM, et al. Pediatric reference intervals for 28 chemistries and immunoassays on the Roche cobas® 6000 analyzer—A CALIPER pilot study. Clin Biochem 2010;43;1045–50.
	6	Ghoshal AK, Soldin SJ. Evaluation of the Dade Behring Dimension RxL: integrated chemistry system-pediatric reference ranges. Clin Chim Acta 2003;331:135–46.
Method(s)	1,2	Arsenazo III dye method. Vitros 700 (Ortho-Clinical Diagnostics, Raritan, NJ).
	3	Cresolphthalein complexone. Hitachi 747 Boehringer Mannheim reagents (Boehringer Mannheim Diagnostics, Indianapolis, IN).
	4	Abbott Architect ci8200 (Abbott Diagnostics, Abbott Park, IL).
	5	Roche cobas® 6000 analyzer (Roche Diagnostics Limited, West Sussex, UK).
	6	The calcium method is a modification of the calcium o-cresolphthalein complexone (OCPC) reaction. Siemens Dimension RxL Analyzer (Siemens Healthcare Diagnostics, Deerfield, IL).
Comment(s)	1,3	Study used hospitalized patients and a computerized approach to removing outliers. Values are 2.5–97.5th percentiles.
	2	From normal healthy children. Values are 2.5–97.5th percentiles. *No significant differences were found for males and females. These ranges were therefore derived from combined data.
	4	1459 serum/plasma specimens from children attending select outpatient clinics were collected from the five age groups noted above. Values are 2.5–97.5th percentiles.
	5	Approximately 600 outpatient samples from a pediatric population deemed to be metabolically stable were subdivided into five age classes ranging from 0 to 20 years of age and further partitioned by gender. Values are 2.5–97.5th percentiles.
	6	**Reference ranges were obtained by comparing results from previously published data and using regression equations.

CALCIUM, IONIZED

Test	Age	n	Male mg/dL	Male mmol/L	Female mg/dL	Female mmol/L
1	0–1 mo	207	3.9–6.0	1.0–1.5	3.9–6.0	1.0–1.5
	1–6 mo	96	3.7–5.9	0.95–1.5	3.7–5.9	0.95–1.5
2	1–19 y	*	4.9–5.5	1.22–1.37		
	20 y–adult		4.75–5.3	1.18–1.32		
	1–17 y				4.9–5.5	1.22–1.37
	18 y–adult				4.75–5.3	1.18–1.32

Specimen Type(s)	1,2	Whole blood
Reference(s)	1	Snell J, Greeley C, Colaco A, et al. Pediatric reference ranges for arterial pH, whole blood electrolytes and glucose. Clin Chem 1993:39;1173. (Abstract)
	2	Burritt MF, Slockbower JM, Forsman RW, et al. Pediatric reference intervals for 19 biologic variables in healthy children. Mayo Clinic Proceedings 1990:65;329–36.
Method(s)	1	Ciba Corning 288 Blood Gas System (Ciba Corning Diagnostics, East Walpole, MA).
	2	Ion selective electrode Radiometer ICA 1 (Radiometer America, Inc., Cleveland, OH).
Comment(s)	1	Study used hospitalized patients and a computerized approach to removing outliers. Values are 2.5–97.5th percentiles.
	2	From normal healthy children. Values are 2.5–97.5th percentiles.
		*See reference for numbers.

CARBON DIOXIDE (CO_2)

Test	Age	n	mmol/L
	Male and Female		
1	0–1 wk	≥100	17–26
	1 wk–1 mo	≥100	17–27
	1–6 mo	≥100	17–29
	6 mo–1 y	≥100	18–29
	1 y	≥100	20–31
2	Infants*	*	13–29
	Adults	**	24–30
3	0–1 wk	***	13–21
	1 wk–1 mo		13–22
	1–6 mo		13–23
	6 mo–1 y		14–23
	>1 y		16–25
4	0–12 mo	124	13.5–22.1
	1–5 y	156	13.7–22.1
	6–10 y	135	14.2–23.2
	11–14 y	153	14.8–24.6
	15–20 y	183	15.5–25.2

Specimen Type(s)	1	Plasma
	2	Cord blood
	3,4	Plasma/serum
Reference(s)	1	Greeley C, Snell J, Colaco A, et al. Pediatric reference ranges for electrolytes and creatinine. Clin Chem 1993;39:1172. (Abstract)
	2	Chemistry analytes in arterial and venous umbilical cord blood. Clin Chem 1993;39:1041–4.
	3	Ghoshal AK, Soldin SJ. Evaluation of the Dade Behring Dimension RxL: integrated chemistry system-pediatric reference ranges. Clin Chim Acta 2003;331:135–46.
	4	Chan MK, Seiden-Long I, Aytekin M, et al. Canadian Laboratory Initiative on Reference Interval Database (CALIPER): pediatric reference intervals for an integrated clinical chemistry and immunoassay analyzer, Abbott ARCHITECT ci8200. Clin Biochem 2009;42:885–91.
Method(s)	1	Vitros 700 (Ortho-Clinical Diagnostics, Raritan, NJ).
	2	Hitachi 737, Boehringer Mannheim reagents (Boehringer Mannheim, Montreal, Canada).
	3	The total carbon dioxide method uses a Severinghaus electrode designed to measure the liberated CO_2 from an acidified sample. Siemens Dimension RxL Analyzer (Siemens Healthcare Diagnostics, Deerfield, IL).
	4	Abbott Architect ci8200 (Abbott Diagnostics, Abbott Park, IL).
Comment(s)	1	Study used hospitalized patients and a computerized approach to removing outliers. Values are 2.5–97.5th percentiles.
	2	*Study used 397 infants (209 girls and 188 boys delivered between 37 and 41 weeks gestation).
		**Numbers not provided.
		Results are 2.5–97.5th percentiles.
	3	***Numbers not provided.
		Reference ranges were obtained by comparing results from previously published data and using regression equations.
	4	1459 serum/plasma specimens from children attending select outpatient clinics were collected from the five age groups noted above. Values are 2.5–97.5th percentiles.

CARBON DIOXIDE, PARTIAL PRESSURE (pCO$_2$)

Test	Age	n	mmHg	kPa
	Male and Female			
1	Newborn	*	27–40	3.6–5.3
	Infant		27–41	3.6–5.5
	Thereafter		32–48	4.3–6.4
2	Premature neonates	248	39–68	5.2–9.1

Specimen Type(s)	1	Arterial whole blood
	2	Capillary blood
Reference(s)	1	Behrman RE, ed. Nelson textbook of pediatrics, 14th ed. Philadelphia, PA: WB Saunders Company, 1992:1818.
	2	Soldin SJ. Children's National Medical Center. Unpublished data.
Method(s)	1	Not given.
	2	I-Stat (Abbott Laboratories, Abbott Park, IL).
Comment(s)	1	*Numbers not provided.
	2	Study used hospitalized patients and a computerized approach adapted from the Hoffmann technique.

		Male		Female	
CARNITINE (TOTAL)					
Test	Age	n	µmol/L	n	µmol/L
1	1–7 d	*	17–46	*	17–46
	2 y		24–66		24–66
	<2 y		37–89		30–73
2	1–12 mo	12	15–39	12	15–39
	1–7 y	27	18–37	27	18–37
	7–15 y	9	31–43	9	31–43
Specimen Type(s)	1	Serum			
	2	Plasma			
Reference(s)	1	Mayo Medical Laboratories test catalog, 1994:78.			
	2	Bonnefont JP, Specola NP, Vassault A, et al. The fasting test in pediatrics: application to the diagnosis of pathological hypo- and hyperketotic states. Eur J Pediatr 1990;150:80–5.			
Method(s)	1	Radioisotope enzymatic.			
	2	Radioisotopic assay. See reference.			
Comment(s)	*Note:*	Free carnitine normally comprises 60–80% of the total carnitine.			
	1	The total carnitine minus the free carnitine gives the esterified carnitine. *Numbers not provided.			
	2	The above results are 10–90th percentiles and refer to 15-h fasting values. For 20- and 24-h fasting values, see reference.			

CERULOPLASMIN

Test	Age	Male n	Male mg/L	Female n	Female mg/L
1	0–5 d	73*	55–286	73*	55–286
	1–3 y	51*	264–506	51*	264–506
	4–6 y	39*	264–462	39*	264–462
	7–9 y	39*	264–440	39*	264–440
	10–13 y	36	242–396	45	253–473
	14–19 y	46	154–374	66	220–495
2	1–30 d	35	77–253	36	33–275
	31–365 d	119	154–484	87	154–429
	1–3 y	127	253–561	114	286–539
	4–6 y	99	286–561	81	264–539
	7–9 y	75	253–517	84	231–484
	10–12 y	69	209–506	90	209–484
	13–15 y	73	198–495	72	209–462
	16–18 y	49	198–451	73	220–495

Specimen Type(s)	1	Serum
	2	Plasma/serum
Reference(s)		The above values have been adjusted from those published to convert results to current IFCC units using IFCC guidelines/standards.
	1	Lockitch G, Halstead AC, Quigley G, et al. Age- and sex-specific pediatric reference intervals: study design and methods illustrated by measurement of serum proteins with the Behring LN nephelometer. Clin Chem 1988;34:1618–21.
	2	Soldin SJ, Hicks JM, Bailey J, et al. Pediatric reference ranges for β2-microglobulin and ceruloplasmin. Clin Chem 1997;43:S199. (Abstract)
Method(s)	1	Nephelometry using Behring antisera and Behring LN nephelometer (Behring Diagnostics, Hoechst Canada, Inc., Montreal, Canada).
	2	Behring nephelometer with Behring reagents (Behring Diagnostics, Westwood, MA).
Comment(s)	1	Healthy normal children. Results represent the 0.025–0.975 fractiles. Males and females had similar ranges from 0–9 y and numbers quoted are combined for these age ranges. *No significant differences were found for males and females. These ranges were therefore derived from combined data.
	2	Study used hospitalized patients and a computerized approach adapted from the Hoffmann technique. Values are the 2.5–97.5th percentiles.

CHLORIDE

		Male and Female	
Test	Age	n	mmol/L
1	0–1 wk	≥100	96–111
	1 wk–1 mo	≥100	96–110
	1–6 mo	≥100	96–110
	6 mo–1 y	≥100	96–108
	>1 y	≥100	96–109
2	1–17 y	*	102–112
	18–Adult		100–108
3	0–7 d	**	97–108
	7–31 d		97–108
	1–6 mo		97–108
	6 mo–1 y		97–106
	>1 y		97–107

Specimen Type(s)	1	Plasma
	2	Serum
	3	Plasma/serum
Reference(s)	1	Greeley C, Snell J, Colaco A, et al. Pediatric reference ranges for electrolytes and creatinine. Clin Chem 1993;39:1172. (Abstract)
	2	Burritt MF, Slockbower JM, Forsman BS, et al. Pediatric reference intervals for 19 biologic variables in healthy children. Mayo Clinic Proceedings 1990;65:329–36.
	3	Ghoshal AK, Soldin SJ. Evaluation of the Dade Behring Dimension RxL: integrated chemistry system-pediatric reference ranges. Clin Chim Acta 2003;331:135–46.
Method(s)	1	Vitros (Ortho-Clinical Diagnostics, Raritan, NJ).
	2	Coulometric-Beckman Astra 8 (Beckman Instruments, Inc., Palo Alto, CA).
	3	The sodium, potassium, and chloride (Na/K/Cl) methods use indirect sample sensing with the QuikLYTE® Integrated Multisensor Technology (IMT) to develop an electrical potential proportional to the activity of each specific ion in the sample. The total carbon dioxide (TCO_2) method uses a Severinghaus electrode designed to measure the liberated CO_2 from an acidified sample. Siemens Dimension RxL Analyzer (Siemens Healthcare Diagnostics, Deerfield, IL).
Comment(s)	1	Study used hospitalized patients and a computerized approach to removing outliers. Values are 2.5–97.5th percentiles. N ≥100 in all of the above.
	2	From normal children healthy children. Values are 2.5–97.5th percentiles. *See reference for numbers.
	3	**Reference ranges were obtained by comparing results from previously published data and using regression equations.

CHOLESTEROL

Test	Age	Male			Female		
		n	mg/dL	mmol/L	n	mg/dL	mmol/L
1	0–1 mo	37	45–177	1.16–4.58	27	63–198	1.63–5.12
	2–6 mo	354	60–197	1.55–5.09	243	66–218	1.71–5.64
	7–12 mo	401	89–208	2.30–5.39	252	74–218	1.91–5.64
2	1–3 y	49*	44–181	1.15–4.70	49*	44–181	1.15–4.70
	4–6 y	38*	108–187	2.80–4.80	38*	108–187	2.80–4.80
	7–9 y	72*	112–247	2.90–6.40	72*	112–247	2.90–6.40
	10–11 y	28	125–230	3.25–5.95	34	127–244	3.30–6.30
	12–13 y	32	127–230	3.30–5.95	40	125–213	3.25–5.55
	14–15 y	39	106–224	2.75–5.80	50	130–213	3.35–5.55
	16–19 y	41	110–220	2.85–5.70	68	106–217	2.75–5.60
3	1–30 d	62	54–151	1.40–3.90	74	62–155	1.60–4.01
	31–182 d	77	81–147	2.09–3.80	75	62–141	1.60–3.65
	183–365 d	53	76–179	1.97–4.63	45	76–216	1.97–5.59
	1–3 y	136	85–182	2.20–4.71	111	108–193	2.79–4.99
	4–6 y	112	110–217	2.84–5.61	113	106–193	2.74–4.99
	7–9 y	124	110–211	2.84–5.46	104	104–210	2.69–5.43
	10–12 y	111	105–223	2.72–5.77	109	105–218	2.72–5.64
	13–15 y	126	91–204	2.35–5.28	105	108–205	2.79–5.30
	16–18 y	112	82–192	2.12–4.97	110	92–234	2.38–6.05
4	0–12 mo	57	47–205	1.23–5.30	71	93–223	2.42–5.77

	Male and Female			
	Age	n	mg/dL	mmol/L
	1–5 y	152	114–220	2.96–5.71
	6–10 y	133	115–221	2.97–5.72
	11–14 y	153	107–226	2.77–5.86
	15–20 y	172	97–237	2.52–6.14

Test	Age	Male and Female			
		n	mg/dL	mmol/L	
5	0–<1 y	97	91–205	2.36–5.32	
	≥1–≤15 y	371	104–227	2.70–5.89	

		Male		Female		
	Age	n	mg/dL	mmol/L	n	mg/dL	mmol/L
	>15–≤20 y	27	95–208	2.46–5.39	42	93–233	2.42–6.04

Test	Age	n	Male		Female	
			mg/dL	mmol/L	mg/dL	mmol/L
6	0–1 mo	**	38–174	0.98–4.50	56–195	1.45–5.04
	2–6 mo		53–194	1.37–5.02	59–216	1.53–5.59
	7–12 mo		83–205	2.15–5.30	68–216	1.76–5.59
	1–3 y		37–178	0.96–4.60	37–178	0.96–4.60
	4–6 y		103–184	2.66–4.76	103–184	2.66–4.76
	7–9 y		107–245	2.77–6.34	107–245	2.77–6.34
	10–11 y		120–228	3.10–5.90	122–242	3.16–6.26
	12–13 y		122–228	3.16–5.90	120–211	3.10–5.46
	14–15 y		101–222	2.61–5.74	125–211	3.23–5.46
	16–18 y		105–218	2.72–5.64	101–215	2.61–5.56

Specimen Type(s)	1–6	Plasma/serum
Reference(s)	1	Soldin SJ, Rakotoarisoa FTS. Pediatric reference ranges for amylase and cholesterol on the Kodak Ektachem 500 in the first year of life. Clin Chem 1996;42:S308. (Abstract)
	2	Lockitch G, Halstead AC, Albersheim S, et al. Age- and sex-specific pediatric reference intervals for biochemistry analytes as measured with the Ektachem-700 analyzer. Clin Chem 1988;34:1622–5.
	3	Hicks JM, Bailey J, Beatey J, et al. Pediatric reference ranges for cholesterol. Clin Chem 1996;42:S307. (Abstract)
	4	Chan MK, Seiden-Long I, Aytekin M, et al. Canadian Laboratory Initiative on Reference Interval Database (CALIPER): pediatric reference intervals for an integrated clinical chemistry and immunoassay analyzer, Abbott ARCHITECT ci8200. Clin Biochem 2009;42:885–91.
	5	Kulasingam V, Jung BP, Blasutig IM, et al. Pediatric reference intervals for 28 chemistries and immunoassays on the Roche cobas® 6000 analyzer—A CALIPER pilot study. Clin Biochem 2010;43;1045–50.
	6	Ghoshal AK, Soldin SJ. Evaluation of the Dade Behring Dimension RxL: integrated chemistry system-pediatric reference ranges. Clin Chim Acta 2003;331:135–46.
Method(s)	1,2	Cholesterol oxidase method. Vitros 700 (Ortho-Clinical Diagnostics, Raritan, NJ).
	3	Boehringer Mannheim reagents on the Hitachi 747 analyzer (Boehringer Mannheim Diagnostics, Indianapolis, IN).
	4	Abbott Architect ci8200 (Abbott Diagnostics, Abbott Park, IL).
	5	Roche cobas® 6000 analyzer (Roche Diagnostics Limited, West Sussex,UK).
	6	Cholesterol esterase catalyzes the hydrolysis of cholesterol esters to produce free cholesterol, which along with preexisting free cholesterol is oxidized in a reaction catalyzed by cholesterol oxidase. Siemens Dimension RxL Analyzer (Siemens Healthcare Diagnostics, Deerfield, IL).
Comment(s)	1,3	Values are 2.5–97.5th percentiles. Study used hospitalized patients and a computerized approach to removing outliers.
	2	The study population was healthy children. Non-parametric methods were used to determine the reference values. The central 95% were used. *Males and females were not studied separately; n refers to the total number of males and females studied in each age group.
	4	1459 serum/plasma specimens from children attending select outpatient clinics were collected from the five age groups noted above. Values are 2.5–97.5th percentiles.
	5	Approximately 600 outpatient samples from a pediatric population deemed to be metabolically stable were subdivided into five age classes ranging from 0 to 20 years of age and further partitioned by gender. Values are 2.5–97.5th percentiles.
	6	**Reference ranges were obtained by comparing results from previously published data and using regression equations.

COENZYME Q10

	Male and Female			
Test	Age	n	pmol/µL	
1	1–16 y (mean 6.3 y)	50	0.38–1.20	
2	10.9–13.3 y (mean 12.6 y)	50	0.62–0.95	

	Male and Female		
Test	Age	n	µmol/µL
3	0.2–7.6 y	50	0.42–1.70
	11–22 y	40	0.50–1.34

Specimen Type(s)	1–3	Plasma
Reference(s)	1	Niklowitz P, Menke T, Andler W, et al. Simultaneous analysis of coenzyme Q10 in plasma, erythrocytes and platelets: comparison of the antioxidant level in blood cells and their environment in healthy children and after oral supplementation in adults. Clin Chim Acta 2004;342:219–26.
	2	Menke T, Niklowitz P, de Sousa G, Reinehr T, Andler W. Comparison of coenzyme Q10 plasma levels in obese and normal weight children. Clin Chim Acta 2004;349:121–7.
	3	Miles MV, Horn PS, Tang PH, Morrison JA, Miles L, DeGrauw T, Pesce AJ. Age-related changes in plasma coenzyme Q10 concentrations and redox state in apparently healthy children and adults. Clin Chim Acta 2004;347:139–44.
Method(s)	1,2	HPLC.
	3	HPLC with electrochemical detection.
Comment(s)	1,2	Values are the 5–95th percentiles.
	3	Values are 2.5–97.5th percentiles.

COMPLEMENT FRACTION — C_{1r}

Male and Female		
Age	n	mg/L
Term newborn cord blood	125	27–65
<18 y	31	25–140

Specimen Type(s)	Plasma (EDTA or citrate)
Reference(s)	Sonntag J, Brandenburg U, Polzehl D, et al. Complement System in healthy term newborns: reference values in umbilical cord blood. Ped Develop Path 1998;1:131–5.
Method(s)	Single radial immunodiffusion (Binding Site, Birmingham, UK).
Comment(s)	Studies were performed on cord blood obtained from 125 healthy term newborns. Results are 5–95th percentiles.

COMPLEMENT FRACTION — C_2

Male and Female		
Age	n	mg/L
Term newborn cord blood	125	12–24
<18 y	31	18–40

Specimen Type(s)	Plasma (EDTA or citrate)
Reference(s)	Sonntag J, Brandenburg U, Polzehl D, et al. Complement System in healthy term newborns: reference values in umbilical cord blood. Ped Develop Path 1998;1:131–5.
Method(s)	Single radial immunodiffusion (Binding Site, Birmingham, UK).
Comment(s)	Studies were performed on cord blood obtained from 125 healthy term newborns. Results are 5–95th percentiles.

COMPLEMENT FRACTION — C_{3a}

Male and Female		
Age	n	µg/L
Term newborn cord blood	125	4–255
<18 y	31	0–161

Specimen Type(s)	Plasma (EDTA or citrate)
Reference(s)	Sonntag J, Brandenburg U, Polzehl D, et al. Complement System in healthy term newborns: reference values in umbilical cord blood. Ped Develop Path 1998;1:131–5.
Method(s)	Enzyme immunoassay (EIA: Fa. Progen Biotechnik GmbH, Heidelberg, Germany).
Comment(s)	Studies were performed on cord blood obtained from 125 healthy term newborns. Results are 5–95th percentiles.

COMPLEMENT FRACTION — C_{3c}

Test	Age	Male			Female		
		n	mg/dL	g/L	n	mg/dL	g/L
1	0–6 mo	70	56–150	0.56–1.50	71	56–135	0.56–1.35
	6 mo–3 y	121	86–179	0.86–1.79	117	72–164	0.72–1.64
	4–6 y	66	101–191	1.01–1.91	54	96–144	0.96–1.44
	7–9 y	70	116–210	1.16–2.10	38	104–186	1.04–1.86
	10–12 y	82	111–161	1.11–1.61	73	110–191	1.10–1.91
	13–15 y	68	98–201	0.98–2.01	74	111–177	1.11–1.77
	16–18 y	36	114–188	1.14–1.88	57	92–180	0.92–1.80

Test	Age		Male		Female	
		n	mg/dL	g/L	mg/dL	g/L
2	0–6 mo	*	38–100	0.38–1.00	39–91	0.39–0.91
	6 mo–3 y		58–119	0.58–1.19	49–109	0.49–1.09
	4–6 y		68–127	0.68–1.27	65–122	0.65–1.22
	7–9 y		78–140	0.78–1.10	70–124	0.70–1.24
	10–12 y		75–107	0.75–1.07	74–127	0.74–1.27
	13–15 y		66–134	0.66–1.34	75–118	0.75–1.18
	16–18 y		77–125	0.77–1.25	62–120	0.62–1.20

Test	Age	Male and Female		
		n	mg/dL	g/L
3	Healthy infants	32	45–147	0.45–1.47
4	0–5 d	73	39–156	0.39–1.56
	1–19 y	334	77–143	0.77–1.43

Specimen Type(s)	1,3	Serum
	2,4	Plasma/serum
Reference(s)	\multicolumn{2}{l}{The above values have been adjusted from those published in references 1, 3, and 4 to convert results to current IFCC units using IFCC guidelines/standards.}
	1	Soldin SJ, Hicks JM, Bailey J, et al. Pediatric reference ranges for complement factors C3c and C4. Clin Chem 1998;44:A14. (Abstract)
	2	Ghoshal AK, Soldin SJ. Evaluation of the Dade Behring Dimension RxL: integrated chemistry system-pediatric reference ranges. Clin Chim Acta 2003;331:135–46.
	3	Zilow G, Zilow EP, Burger R, et al. Complement activation in newborn infants with early onset infection. Ped Res 1993;34:199–203.
	4	Lockitch G, Halstead AC, Quigley G, et al. Age- and sex-specific pediatric reference intervals: study design and methods illustrated by measurement of serum proteins with the Behring LN nephelometer. Clin Chem 1988;34:1618–21.
Method(s)	1	Behring nephelometer using Behring kits (Behring Diagnostics Inc. Westwood, MA).
	2	The C_3 method is a quantitative turbidimetric assay using endpoint detection based on the precipitation of C_3 by its polyclonal antibody. C_3 from serum reacts with its polyclonal antibody to form immune complexes. Addition of polyethylene glycol accelerates the formation of these complexes. The resulting turbidity is proportional to the concentration of C_3 in the sample. Siemens Dimension RxL Analyzer (Siemens Healthcare Diagnostics, Deerfield, IL).
	3	Radial Immunodiffusion (Behring Diagnostics, Frankfurt, Germany).
	4	Nephelometric on Behring LN nephelometer (Behring Diagnostics, Hoechst Canada, Inc., Montreal).
Comment(s)	1	The study used serum/plasma from hospitalized patients and employed Chauvenet's criteria for removing outliers and a computerized approach adapted from the Hoffmann technique to obtain the 2.5–97.5th percentiles.
	2	*Reference ranges were obtained by comparing results from previously published data and using regression equations.
	3	C_3 reported in serum from normal neonates. Results are 0–100th percentiles.
	4	Healthy children and adults. Results are 0.025–0.975 fractiles.

COMPLEMENT FRACTION — C_4

Test	Age	Male and Female		
		n	mg/dL	g/L
1	0–5 d	73	5–33	0.05–0.33
	1–19 y	334	7–40	0.07–0.40
	Adult	30	10–37	0.10–0.37

Test	Age	Male			Female		
		n	mg/dL	g/L	n	mg/dL	g/L
2	0–6 mo	68	9–28	0.09–0.28	71	10–25	0.10–0.25
	6 mo–3 y	121	14–48	0.14–0.48	116	14–42	0.14–0.42
	4–6 y	66	20–47	0.20–0.47	54	17–41	0.17–0.41
	7–9 y	69	17–35	0.17–0.35	38	16–41	0.16–0.41
	10–12 y	82	18–36	0.18–0.36	73	14–47	0.14–0.47
	13–15 y	68	17–51	0.17–0.51	74	16–37	0.16–0.37
	16–18 y	36	17–38	0.17–0.38	57	11–40	0.11–0.40

Test	Age		Male		Female	
		n	mg/dL	g/L	mg/dL	g/L
2	0–6 mo	*	13–30	0.13–0.30	14–28	0.14–0.28
	6 mo–3 y		17–48	0.17–0.48	17–43	0.17–0.43
	4–6 y		23–47	0.23–0.47	21–42	0.21–0.42
	7–9 y		21–37	0.21–0.37	20–42	0.20–0.42
	10–12 y		21–38	0.21–0.38	18–47	0.18–0.47
	13–15 y		21–51	0.21–0.51	20–38	0.20–0.38
	16–18 y		21–39	0.21–0.39	15–41	0.15–0.41

Specimen Type(s)	1	Serum
	2,3	Plasma/serum
Reference(s)	\multicolumn{2}{l	}{The above values have been adjusted from those published in references 1 and 2 to convert results to current IFCC units using IFCC guidelines/standards.}
	1	Lockitch G, Halstead AC, Quigley G, et al. Age- and sex-specific pediatric reference intervals: study design and methods illustrated by measurement of serum proteins with the Behring LN nephelometer. Clin Chem 1988;34:1618–21.
	2	Soldin SJ, Hicks JM, Bailey J, et al. Pediatric reference ranges for complement factors C3c and C4. Clin Chem 1998;44:A14. (Abstract)
	3	Ghoshal AK, Soldin SJ. Evaluation of the Dade Behring Dimension RxL: integrated chemistry system-pediatric reference ranges. Clin Chim Acta 2003;331:135–46.
Method(s)	1,2	Nephelometric, Behring LN nephelometer (Behring Diagnostics, Hoechst Canada, Inc., Montreal, and Westwood, MA).
	3	The C_4 method is a quantitative turbidimetric assay using endpoint detection based on the precipitation of C_4 by its polyclonal antibody. Siemens Dimension RxL Analyzer (Siemens Healthcare Diagnostics, Deerfield, IL).
Comment(s)	1	Normal healthy children. Values provided are 0.025–0.975 fractiles.
	2	The study used serum/plasma from hospitalized patients and employed Chauvenet's criteria for removing outliers and a computerized approach adapted from the Hoffmann technique to obtain the 2.5–97.5th percentiles.
	3	*Reference ranges were obtained by comparing results from previously published data and using regression equations.

COMPLEMENT FRACTION — C_5

Male and Female

Age	n	mg/L
Term newborn cord blood	125	64–127
Adult range	31	83–169

Specimen Type(s)	Plasma (EDTA or citrate)
Reference(s)	Sonntag J, Brandenburg U, Polzehl D, et al. Complement System in healthy term newborns: reference values in umbilical cord blood. Ped Develop Path 1998;1:131–5.
Method(s)	Single radial immunodiffusion (Binding Site, Birmingham, UK).
Comment(s)	Studies were performed on cord blood obtained from 125 healthy term newborns. Results are 5–95th percentiles.

COMPLEMENT FRACTION — C_{5a}

Male and Female

Age	n	µg/L
Term newborn cord blood	125	0.11–1.19
Adult range	31	0.00–0.62

Specimen Type(s)	Plasma (EDTA or citrate)
Reference(s)	Sonntag J, Brandenburg U, Polzehl D, et al. Complement System in healthy term newborns: reference values in umbilical cord blood. Ped Develop Path 1998;1:131–5.
Method(s)	Specific Sandwich EIA (FA. Behring, Marburg, Germany).
Comment(s)	Studies were performed on cord blood obtained from 125 healthy term newborns. Results are 5–95th percentiles.

COMPLEMENT FRACTION — C_7

Male and Female

Age	n	mg/L
Term newborn cord blood	125	32–89
Adult range	31	35–79

Specimen Type(s)	Plasma (EDTA or citrate)
Reference(s)	Sonntag J, Brandenburg U, Polzehl D, et al. Complement System in healthy term newborns: reference values in umbilical cord blood. Ped Develop Path 1998;1:131–5.
Method(s)	Single radial immunodiffusion (Binding Site, Birmingham, UK).
Comment(s)	Studies were performed on cord blood obtained from 125 healthy term newborns. Results are 5–95th percentiles.

COMPLEMENT FRACTION — FACTOR D

Male and Female

Age	n	mg/L
Term newborn cord blood	125	3.6–7.3
Adult range	31	2.7–5.4

Specimen Type(s)	Plasma (EDTA or citrate)
Reference(s)	Sonntag J, Brandenburg U, Polzehl D, et al. Complement System in healthy term newborns: reference values in umbilical cord blood. Ped Develop Path 1998;1:131–5.
Method(s)	Single radial immunodiffusion (Binding Site, Birmingham, UK).
Comment(s)	Studies were performed on cord blood obtained from 125 healthy term newborns. Results are 5–95th percentiles.

COMPLEMENT FRACTION — FACTOR H

Male and Female		
Age	n	mg/L
Term newborn cord blood	125	178–296
Adult range	31	146–553

Specimen Type(s)	Plasma (EDTA or citrate)
Reference(s)	Sonntag J, Brandenburg U, Polzehl D, et al. Complement System in healthy term newborns: reference values in umbilical cord blood. Ped Develop Path 1998;1:131–5.
Method(s)	Single radial immunodiffusion (Binding Site, Birmingham, UK).
Comment(s)	Studies were performed on cord blood obtained from 125 healthy term newborns. Results are 5–95th percentiles.

COMPLEMENT FRACTION — FACTOR I

Male and Female		
Age	n	mg/L
Term newborn cord blood	125	15–32
Adult range	31	32–88

Specimen Type(s)	Plasma (EDTA or citrate)
Reference(s)	Sonntag J, Brandenburg U, Polzehl D, et al. Complement System in healthy term newborns: reference values in umbilical cord blood. Ped Develop Path 1998;1:131–5.
Method(s)	Single radial immunodiffusion (Binding Site, Birmingham, UK).
Comment(s)	Studies were performed on cord blood obtained from 125 healthy term newborns. Results are 5–95th percentiles.

COMPLEMENT FRACTION — PROPERDIN

Male and Female		
Age	n	mg/L
Term newborn cord blood	125	5.6–14.2
Adult range	31	24–50

Specimen Type(s)	Plasma (EDTA or citrate)
Reference(s)	Sonntag J, Brandenburg U, Polzehl D, et al. Complement System in healthy term newborns: reference values in umbilical cord blood. Ped Develop Path 1998;1:131–5.
Method(s)	Single radial immunodiffusion (Binding Site, Birmingham, UK).
Comment(s)	Studies were performed on cord blood obtained from 125 healthy term newborns. Results are 5–95th percentiles.

COPPER

Test	Age	Male			Female		
		n	µg/dL	µmol/L	n	µg/dL	µmol/L
1	0–5 d	27*	9–46	1.4–7.20	27*	9–46	1.4–7.2
	1–5 y	77*	80–150	12.6–23.6	77*	80–150	12.6–23.6
	6–9 y	44*	84–136	13.2–21.4	44*	84–136	13.2–21.4
	10–14 y	36	80–121	12.6–19.0	23	82–120	12.9–18.9
	15–19 y	55	64–171	10.1–18.4	31	72–160	11.3–25.2

Test	Age	Male and Female		
		n	µg/dL	µmol/L
2	0–<0.5 y	13	38–104	5.9–16.3
	0.5–<1.0 y	18	24–152	3.8–23.8
	1.0–<2.0 y	15	76–193	11.9–30.3
	2.0–<4.0 y	23	87–187	13.7–29.3
	4.0–<6.0 y	19	56–191	8.8–30.0
	6.0–<10.0 y	25	117–181	18.4–28.4
	10.0–<14.0 y	21	87–182	13.7–28.5
	14.0–<18.0 y	17	75–187	11.7–29.3

Specimen Type(s)	1	Serum
	2	Plasma/serum
Reference(s)	1	Lockitch G, Halstead A, Wadsworth L, et al. Age- and sex-specific pediatric reference intervals for zinc, copper, selenium, iron, vitamins A and E, and related proteins. Clin Chem 1988;34:1625–8.
	2	Rükgauer M, Klein J, Kruse-Jarres JD. Reference values for the trace elements copper, manganese, selenium, and zinc in the serum/plasma of children, adolescents, and adults. J Trace Elements Med Biol 1997;11:92–8.
Method(s)	1	Electrothermal atomic absorption spectroscopy with deuterium background correction. Varian GTA-95 (Varian Canada, Inc., Mississauga, Canada).
	2	Atomic absorption spectrophotometry with Zeeman background compensation. Perkin Elmer ETAAS, Zeeman 3030 (Uberlingen, Germany).
Comment(s)	1	The study population was healthy children. Non-parametric methods were used to determine the 0.025 and 0.975 fractiles. *No significant differences were found for males and females. These ranges were therefore derived from combined data.
	2	Study population was drawn from patients visiting the outpatient department or surgical or orthopedic ward for preoperative workup. Results are mean ±2 SDs (2.5–97.5th percentiles).

CORTISOL

Test	Age	\multicolumn{3}{c}{5–11 am}	\multicolumn{3}{c}{5–11 pm}

		\multicolumn{6}{c}{Male and Female}
Test	Age	n	µg/L	nmol/L	n	µg/L	nmol/L
1	0–24 mo	33	10–340	28–938	25	10–300	28–828
	2–<11 y	31	10–330	28–911	23	10–240	28–662
	11–18 y	31	10–280	28–773	19	10–220	28–607

		\multicolumn{4}{c}{Male and Female}
Test	Age	n	µg/L	nmol/L
2	0–<24 mo	149	<10–350	<28–966
	2–<6 y	47	<10–260	<28–717
	6–<11 y	64	<10–380	<28–1049
	11–<15 y	71	20–250	55–690
	15–<19 y	45	<10–310	<28–856
3	0–12 mo	90	5–159	15–439
	1–5 y	145	17–145	47–399
	6–10 y	126	13–149	37–410
	11–14 y	164	12–148	34–408

		Male			Female		
	Age	n	µg/L	nmol/L	n	µg/L	nmol/L
	15–20 y	37	15–148	42–409	95	25–234	68–644
4	5 d	*	6–198	17–546	*	6–198	17–546
	2–12 mo		24–229	66–632		24–229	66–632
	2–13 y		25–229	69–632		25–229	69–632
	14–15 y		25–229	69–632		24–286	66–789
	16–18 y		24–286	66–789		24–286	66–789
5	5 d	**	6–156	16–431	**	6–156	16–431
	2–12 mo		20–181	54–499		20–181	54–499
	2–13 y		21–181	57–499		21–181	57–499
	14–15 y		21–181	57–499		20–274	54–756
	16–18 y		20–225	54–622		20–274	54–756

		\multicolumn{4}{c}{Male and Female}
Test	Age	n	µg/L	nmol/L
6	0–≤15 y	419	21–205	58–567

		Male			Female		
	Age	n	µg/L	nmol/L	n	µg/L	nmol/L
	>15–≤20 y	23	37–293	102–807	29	17–132	47–363

Specimen Type(s)	1,4,5	Serum
	2,3,6	Plasma/serum
Reference(s)	1	Soldin SJ, Murthy JN, Agarwalla PK, et. al. Pediatric reference ranges for creatine kinase, CKMB, troponin I, iron and cortisol. Clin Biochem 1999; 32:77–80.
	2	Soldin OP, Hoffman EG, Waring MA, Soldin SJ. Pediatric reference intervals for FSH, LH, estradiol, T3, free T3, cortisol, and growth hormone on the DPC IMMULITE 1000. Clin Chim Acta 2005;355:205–10.
	3	Chan MK, Seiden-Long I, Aytekin M, et al. Canadian Laboratory Initiative on Reference Interval Database (CALIPER): pediatric reference intervals for an integrated clinical chemistry and immunoassay analyzer, Abbott ARCHITECT ci8200. Clin Biochem 2009;42:885–91.
	4	Jonetz-Mentzel L, Wiedemann G. Establishment of reference ranges for cortisol in neonates, infants, children and adolescents. Eur J Clin Chem Biochem 1993;31:525–9.
	5	Murthy JN, Hicks JM, Soldin SJ. Evaluation of the Technicon Immuno I Random Access Immunoassay Analyzer and calculation of pediatric reference ranges for endocrine tests, T-uptake, and ferritin. Clin Biochem 1995;28:181–5.
	6	Kulasingam V, Jung BP, Blasutig IM, et al. Pediatric reference intervals for 28 chemistries and immunoassays on the Roche cobas® 6000 analyzer—A CALIPER pilot study. Clin Biochem 2010;43;1045–50.
Method(s)	1	Bayer Immuno I using Bayer reagents (Bayer Corp., Tarrytown, NY).
	2	Siemens IMMULITE® 1000 analyzer (Siemens Healthcare Diagnostics, Deerfield, IL).
	3	Abbott Architect ci8200 (Abbott Diagnostics, Abbott Park, IL).
	4	TDx fluorescence polarization immunoassay (Abbott Laboratories, Abbott Park, IL).
	5	Bayer (Technicon) Immuno I analyzer with Bayer kits (Bayer Corp., Tarrytown, NY).
	6	Roche cobas® 6000 analyzer (Roche Diagnostics Limited, West Sussex, UK).
Comment(s)	1	Study used hospitalized patients (and consequently results are higher) and a computerized approach adapted from the Hoffmann technique. Values are the 2.5–97.5th percentiles. Note that pm values are only moderately lower than am values.
	2	Study used hospitalized patients (and consequently results are higher) and a computerized approach adapted from the Hoffmann technique. Values are the 2.5–97.5th percentiles.
	3	1459 serum/plasma specimens from children attending select outpatient clinics were collected from the five age groups noted above. Values are 2.5–97.5th percentiles (rounded to the nearest whole number).
	4	687 normal healthy neonates, infants, children and adolescents. Values are 2.5–97.5th percentiles. Results were rounded off to the nearest whole integer. *See reference for numbers.
	5	Study obtained from hospitalized patients, and consequently results are higher. Results represent 2.5–97.5th percentiles. **Numbers not provided.
	6	Approximately 600 outpatient samples from a pediatric population deemed to be metabolically stable were subdivided into five age classes ranging from 0 to 20 years of age and further partitioned by gender. Values are 2.5–97.5th percentiles (rounded to the nearest whole number).

C-REACTIVE PROTEIN (CRP)

		Male and Female	
Test	Age	n	mg/L
1	Cord serum	*	0.010–0.35
	Adult		0.068–8.2
2	Cord serum	48	0.015–6.0

		Male			Female		
Test	Age	n	mg/dL	mg/L	n	mg/dL	mg/L
3	0–90 d	47	0.08–1.58	0.8–15.8	39	0.09–1.58	0.9–15.8
	91 d–12 mo	57	0.08–1.12	0.8–11.2	34	0.05–0.79	0.5–7.9
	13–36 mo	67	0.08–1.12	0.8–11.2	52	0.08–0.79	0.8–7.9
	4–10 y	65	0.06–0.79	0.6–7.9	62	0.05–1.00	0.5–10.0
	11–14 y	50	0.08–0.76	0.8–7.6	45	0.06–0.81	0.6–8.1
	15–18 y	39	0.04–0.79	0.4–7.9	42	0.06–0.79	0.6–7.9

Specimen Type(s)	1,2	Serum
	3	Plasma/serum
Reference(s)	1	Burtis CA, Ashwood CR, eds. Tietz textbok of clinical chemistry. 2nd ed. WB Saunders Co., 1994; 2184.
	2	Shine B, Gould J, Campbell C, et al. Serum C-reactive protein in normal and infected neonates. Clin Chim Acta 1985;148:97–103.
	3	Soldin OP, Bierbower LH, Choi JJ, et al. Serum iron, ferritin, transferrin, total iron binding capacity, hs-CRP, LDL cholesterol and magnesium in children; new reference intervals using the Dade Dimension Clinical Chemistry System. Clin Chim Acta 2004;342:211–7.
Method(s)	1	Nephelometric.
	2	Double antibody magnetizable particle radioimmunoassay (1). 1. Shine B, DeBeer FC, Pepys MB. Solid phase radioimmunoassay for human C-reactive protein. Clin Chim Acta 1981;117:13–23.
	3	The CRP method is based on a particle-enhanced turbidimetric immunoassay (PETIA) technique. Siemens Dimension RxL Analyzer (Siemens Healthcare Diagnostics, Deerfield, IL).
Comment(s)	1	*Numbers not provided.
	2	48 cord sera obtained from cord blood of normal neonates. Results are 5–95th percentiles. Values above 10,000 µg/L were found in neonates with infection.
	3	Study used hospitalized patients and a computerized approach adapted from the Hoffmann technique to obtain the 2.5–97.5th percentiles.

CREATINE KINASE

Test	Age	Male n	Male U/L	Female n	Female U/L
1	0–90 d	71	28–300	65	42–470
	3–12 mo	129	24–170	90	26–240
	13–24 mo	121	27–160	120	24–175
	2–10 y	245	30–150	231	24–175
	11–14 y	86	30–150	77	30–170
	15–18 y	56	33–145	111	27–140
2	1–30 d	113	2–183	77	2–134
	31–182 d	95	2–129	64	2–146
	183–365 d	58	2–143	49	18–138
	1–3 y	129	2–163	143	2–134
	4–6 y	133	18–158	112	8–147
	7–9 y	142	2–177	113	26–145
	10–12 y	148	6–217	101	6–137
	13–15 y	144	2–251	124	2–143
	16–18 y	114	2–238	146	13–144

Test	Age	n	Male U/L	Female U/L
3	0–90 d	*	29–303	43–474
	3–12 mo		25–172	27–242
	13–24 mo		28–162	25–177
	2–10 y		31–152	25–177
	11–14 y		31–152	31–172
	15–18 y		34–147	28–142

Specimen Type(s)	1–3	Plasma/serum
Reference(s)	1	Soldin SJ, Murthy JN, Agarwalla PK, et al. Pediatric reference ranges for creatine kinase, CKMB, troponin I, iron and cortisol. Clin Biochem 1999;32:77–80.
	2	Soldin SJ, Hicks JM, Bailey J, et al. Pediatric reference ranges for creatine kinase and insulin-like growth factor 1. Clin Chem 1997;43:S199. (Abstract)
	3	Ghoshal AK, Soldin SJ. Evaluation of the Dade Behring Dimension RxL: integrated chemistry system-pediatric reference ranges. Clin Chim Acta 2003;331:135–46.
Method(s)	1	Vitros 500 (Ortho-Clinical Diagnostics, Raritan, NJ). Creatine phosphate to creatine.
	2	Hitachi 747 using Boehringer Mannheim reagents (Boehringer Mannheim Diagnostics, Indianapolis, IN).
	3	The ATP formed is used to phosphorylate glucose in a reaction catalyzed by hexokinase, and the resulting glucose-6-phosphate is oxidized by glucose-6-phosphate dehydrogenase with the simultaneous reduction of nicotinamide adenine dinucleotide phosphate. Siemens Dimension RxL Analyzer (Siemens Healthcare Diagnostics, Deerfield, IL).
Comment(s)	1,2	Study used hospitalized patients and a computerized adaptation of the Hoffmann technique. Values are 2.5–97.5th percentiles.
	3	*Reference ranges were obtained by comparing results from previously published data and using regression equations.

CREATINE KINASE ISOENZYMES (CPK ISOENZYMES)

Test	Age	n	CKMB %	CKBB %
		Male and Female		
1	Cord blood	*		
	5–8 h		0.3–3.1	0.3–10.5
	24–33 h		1.7–7.9	3.6–13.4
	72–100 h		1.8–5.0	2.3–8.6
	Adult		1.4–5.4	5.1–13.3

Test	Age	n	CKMB 97.5th percentile µg/L
2	0–30 d	76	4.5
	31–90 d	45	4.8
	3–6 mo	88	1.9
	7–12 mo	47	1.7
	1–18 y	97	1.7
3	0–30 d	**	4.2
	31–90 d		4.5
	3–6 mo		1.8
	7–12 mo		1.7
	1–18 y		1.7

Specimen Type(s)	1	Serum
	2,3	Plasma/serum
Reference(s)	1	Behrman RE, ed. Nelson textbook of pediatrics, 14th ed. Philadelphia, PA: WB Saunders Company, 1992:1807.
	2	Soldin SJ, Murthy JN, Agarwalla PK, et al. Pediatric reference ranges for creatine kinase, CKMB, troponin I, iron and cortisol. Clin Biochem 1999;32:77–80.
	3	Ghoshal AK, Soldin SJ. Evaluation of the Dade Behring Dimension RxL: integrated chemistry system-pediatric reference ranges. Clin Chim Acta 2003;331:135–46.
Method(s)	1	Not provided.
	2	Bayer Immuno I with Bayer reagents (Bayer Corp., Tarrytown, NY).
	3	A one-step enzyme immunoassay based on the "sandwich" principle. Siemens Dimension RxL Analyzer (Siemens Healthcare Diagnostics, Deerfield, IL).
Comment(s)	1	*Numbers not provided.
	2	Study used hospitalized patients and a computerized approach adapted from the Hoffmann technique. Values are the 97.5th percentiles.
	3	**Reference ranges were obtained by comparing results from previously published data and using regression equations.

Test	Age	n	mg/dL	µmol/L			
	CREATININE						
	Male and Female						
1	0–1 wk	*	0.6–1.1	53–97			
	1 wk–1 mo		0.3–0.7	27–62			
	1–6 mo		0.2–0.4	18–35			
	7–12 mo		0.2–0.4	18–35			
	1–18 y		0.2–0.7	18–62			
2	0–1 wk	**	0.7–1.2	60–110			
	1 wk–1 mo		0.3–0.8	30–70			
	1 mo–1 y		0.2–0.5	20–40			
	1–9 y		0.2–0.8	20–70			
	10–18 y		0.5–1.1	40–100			
3	0–<1 y	94	0.2–0.4	15–37			
	≥1–≤5 y	136	0.2–0.5	17–43			
	>5–≤10 y	99	0.3–0.6	23–54			
	>10–≤15 y	152	0.4–0.9	31–76			
4	0–1 y	92	0.3–0.5	30–48			
	1–5 y	147	0.4–0.6	31–52			

Test	Age	Male			Female		
		n	mg/dL	µmol/L	n	mg/dL	µmol/L
	6–10 y	53	0.5–0.7	46–63	73	0.4–0.7	36–60
	11–14 y	67	0.5–0.8	46–72	92	0.5–0.8	46–70
	15–20 y	37	0.6–1.0	54–92	110	0.6–0.9	51–81
5	1–30 d	42	0.5–1.2	44–106	40	0.5–0.9	44–80
	31–365 d	62	0.4–0.7	35–62	59	0.4–0.6	35–53
	1–3 y	103	0.4–0.7	35–62	126	0.4–0.7	35–62
	4–6 y	129	0.5–0.8	44–71	116	0.5–0.8	44–71
	7–9 y	121	0.6–0.9	53–80	110	0.5–0.9	44–80
	10–12 y	125	0.6–1.0	53–88	117	0.6–1.0	53–88
	13–15 y	135	0.6–1.2	53–106	141	0.7–1.1	62–97
	16–18 y	106	0.8–1.4	71–123	114	0.8–1.2	71–106

Specimen Type(s)	1,4	Plasma
	2,3,5	Plasma/serum
Reference(s)	1	Greeley C, Snell J, Colaco A, et al. Pediatric reference ranges for electrolytes and creatinine. Clin Chem 1993:39:1172. (Abstract)
	2	Ghoshal AK, Soldin SJ. Evaluation of the Dade Behring Dimension RxL: integrated chemistry system-pediatric reference ranges. Clin Chim Acta 2003;331:135–46.
	3	Kulasingam V, Jung BP, Blasutig IM, et al. Pediatric reference intervals for 28 chemistries and immunoassays on the Roche cobas® 6000 analyzer—A CALIPER pilot study. Clin Biochem 2010;43;1045–50.
	4	Chan MK, Seiden-Long I, Aytekin M, et al. Canadian Laboratory Initiative on Reference Interval Database (CALIPER): pediatric reference intervals for an integrated clinical chemistry and immunoassay analyzer, Abbott ARCHITECT ci8200. Clin Biochem 2009;42:885–91.
	5	Soldin SJ, Hicks JM, Bailey J, et al. Pediatric reference ranges for creatinine on the Hitachi 747 analyzer. Clin Chem 1997;43:S198. (Abstract)
Method(s)	1	Vitros (Ortho-Clinical Diagnostics, Raritan, NJ).
	2	Abbott Architect ci8200 (Abbott Diagnostics, Abbott Park, IL).
	3	Roche cobas® 6000 analyzer (Roche Diagnostics Limited, West Sussex,UK).
	4	Jaffe method on Hitachi 747 Boehringer Mannheim reagents (Boehringer Mannheim Diagnostics, Indianapolis, IN).
	5	The creatinine method employs a modification of the kinetic Jaffe reaction reported by Larsen. Siemens Dimension RxL Analyzer (Siemens Healthcare Diagnostics, Deerfield IL).
Comment(s)	1	Study used hospitalized patients and a computerized approached to removing outliers. *n ≥100. Values are 2.5–97.5th percentiles.
	2	**Reference ranges were obtained by comparing results from previously published data and using regression equations.
	3	Approximately 600 outpatient samples from a pediatric population deemed to be metabolically stable were subdivided into five age classes ranging from 0 to 20 years of age and further partitioned by gender. Values are 2.5–97.5th percentiles.
	4	1459 serum/plasma specimens from children attending select outpatient clinics were collected from the five age groups noted above. Values are 2.5–97.5th percentiles.
	5	Study used hospitalized patients and a computerized approach to removing outliers. Values reported are 2.5–97.5th percentiles.

CREATININE (URINE)

	Male and Female	
Age	n	g/24h
3–8 y	71	0.11–0.68
9–12 y	45	0.17–1.41
13–17 y	42	0.29–1.87
Adult	104	0.63–2.50
Specimen Type(s)	Urine	
Reference(s)	Pediatric endocrine testing. Nichols Institute, 1993:35.	
Method(s)	Kinetic, alkaline picrate.	
Comment(s)	Results are 2.5–97.5th percentiles.	

CYSTATIN C

	Male and Female		
Test	Age	n	mg/L
1	0–3 mo	*	0.8–2.3
	4–12 mo		0.7–1.5
	1–3 y		0.5–1.3
	4–8 y		0.5–1.3
	9–17 y		0.5–1.3
2	<1 mo	12	1.1–2.2
	1–12 mo	29	0.5–1.4
	>12 mo	96	0.5–1.0
3	Preterm infants	58	0.8–2.3
	Full-term infants	58	0.7–1.5
	8 d–1 y	299	0.5–1.3
	>1–3 y	299	0.5–1.3
	>3–16 y	299	0.5–1.3

Specimen Type(s)	1,2	Serum
	3	Plasma/serum
Reference(s)	1	Finney H, Newman DJ, Thakkar H, et al. Reference ranges for plasma cystatin C and creatinine measurements in premature infants, neonates, and older children. Arch Dis Child 2000;82:71–5.
	2	Randers E, Krue S, Erlandsen EJ, Danielsen H, Hansen LG. Reference interval for serum cystatin C in children. Clin Chem 1999;45:1856–8.
	3	Harmoinen A, Ylinen E, Ala-Houhala M, Janas M, Kaila M, Kouri T. Reference intervals for cystatin C in pre- and full-term infants and children. Pediatr Nephrol 2000;15:105–8.
Method(s)	1	Cystatin C is quantified by a specific latex particle-enhanced immunoephelometric assay.
	2	Siemens Nephelometer II (Siemens Healthcare Diagnostics, Deerfield, IL).
	3	Immunoturbidimetric method using the Hitachi 704 analyzer (Boehringer Mannheim Diagnostics, Indianapolis, IN).
Comment(s)	1	*Numbers not provided.
	2,3	Values are 2.5–97.5th percentiles.

DEHYDROEPIANDROSTERONE (DHEA)

Test	Age	Male n	Male ng/dL	Male nmol/L	Female n	Female ng/dL	Female nmol/L
1	6–9 y	15	13–187	0.5–6.5	11	18–189	0.6–6.6
	10–11 y	17	31–205	1.1–7.1	6	112–224	3.9–7.8
	12–14 y	16	83–258	2.9–8.9	12	98–360	3.4–12.5
	Adults	*	180–1250	6.2–43.3	*	130–980	4.5–34
2	0–1 d	**	320–1100	11–39	**	460–1200	16–42
	1–<7 d		90–870	3–30		120–930	4–32
	7–28 d		45–580	1.5–20		90–580	3–20
	1–12 mo		9–290	0.3–10		17–170	0.6–6
	1–<4 y		12–90	0.4–3		20–45	0.7–1.6
	4–10 y						
	P_1		25–300	0.9–10		12–200	0.4–7.0
	P_2		50–580	1.8–20		60–1700	2–60
	P_3		130–640	4.5–22		125–1900	4.4–65
	P_4		190–730	6.5–25		170–1700	6–60
	P_5		230–730	8.0–25		220–810	7.5–28

Male and Female

Test	Age	n	ng/dL	nmol/L
3	0–18 y	122	10–565	0.3–20
4	Premature infants	***		
	26–28 w, day 4		236–3640	8–127
	31–35 w, day 4		80–3150	3–110
	Full-term infants			
	3 d		65–1250	2–44
	7–30 d[a]		50–760	2–27
	1–6 mo		26–385	1–13
	6–12 mo		20–100	1–3.5
	Prepubertal children			
	1–6 y		20–130	1–4.5
	6–8 y[b]		20–275	1–9.6
	8–10 y		31–345	1–12

Test	Male				Female[c]			
	Age	n	ng/dL	nmol/L	Age	n	ng/dL	nmol/L
4 cont.	Puberty Tanner Stage[d]	***			Puberty Tanner Stage[d]	***		
	1 (<9.8 y)		31–345	1–12	1 (<9.2 y)		31–345	1–12
	2 (9.8–14.5 y)		110–495	4–17	2 (9.2–13.7 y)		150–570	5–20
	3 (10.7–15.4 y)		170–585	6–20	3 (10.0–14.4 y)		200–600	7–21
	4 (11.8–16.2 y)		160–640	6–22	4 (10.7–15.6 y)		200–780	7–27
	5 (12.8–17.3 y)		250–900	9–32	5 (11.8–18.6 y)		215–850	7–30
Specimen Type(s)	1,3,4	Serum						
	2	Plasma/serum						
Reference(s)	1	Pediatric endocrine testing. Nichols Institute, 1993:13.						
	2	Soldin SJ, Rifai N, Hicks JM, eds. Biochemical basis of pediatric disease, 3rd ed. Washington, DC: AACC Press, 1998:233.						
	3	Soldin OP, Sharma H, Husted L, Soldin SJ. Pediatric reference intervals for aldosterone, 17alpha-hydroxyprogesterone, dehydroepoandrosterone, testosterone and 25-hydroxy vitamin D3 using tandem mass spectrometry. Clin Biochem 2009;42:823–7.						
	4	Endocrinology expected values. Esoterix Endocrinology, Calabasas Hills, CA. ©2002 Esoterix, Inc., http://www.esoterix.com						
		[a]Data adapted from De Peretti E, Forest MG. Unconjugated dehydroepiandrosterone plasma levels in normal subjects from birth to adolescence in humans: the use of a sensitive radioimmunoassay. J Clin Endocrinol Metab 1976;43:982.						
		[d]Tanner JM, Whitehouse RH. Clinical longitudinal standards for height, weight, height velocity, and the stages of puberty. Arch Dis Childhood 1976;51:170.						
Method(s)	1	Extraction, chromatography, radioimmunoassay.						
	2	See reference.						
	3	Samples were analyzed using isotope dilution liquid chromatography tandem mass spectrometry (LC/MS/MS).						
	4	Not provided.						
Comment(s)	1	Results are 2.5–97.5th percentiles. *Numbers not provided.						
	2	**Numbers not provided. These reference ranges are for guidance only. P_1–P_5 refer to pubertal stages.						
	3	Reference intervals were determined for neonates and children 0–18 years of age. The study was conducted using outpatient samples obtained between January 1, 2004, and June 30, 2008. Values are the 2.5–97.5th percentiles.						
	4	***Numbers not provided.						
		[b]Values begin to increase progressively at about six years of age prior to any physical evidence of puberty.						
		[c]Day of cycle not determined for prepubertal females.						
		[d]As used herein, Tanner stages in males encompass development of both pubic hair and genitalia. In females, each stage encompasses both pubic hair and breast development. While this expands the chronological age range for each stage of pubertal development, it results in a better correlation with hormonal values.						

DEHYDROEPIANDROSTERONE SULFATE (DHEAS)

Test	Age	Male n	Male μg/dL	Male μmol/L	Female n	Female μg/dL	Female μmol/L
1	0–1 mo	56	9–316	0.2–8.6	41	15–261	0.4–7.1
	1–6 mo	69	3–58	0.1–1.6	55	<2–74	<0.1–2.0
	7–12 mo	40	<2–26	<0.1–0.7	28	<2–26	<0.1–0.7
	1–3 y	114	<2–15	<0.1–0.4	121	<2–22	<0.1–0.6
	4–6 y	95	<2–27	<0.1–0.7	86	<2–34	<0.1–0.9
	7–9 y	103	<2–60	<0.1–1.6	83	<2–74	<0.1–2.0
	10–12 y	95	5–137	0.1–3.7	55	3–111	0.1–3.0
	13–15 y	86	3–188	0.1–5.1	78	4–171	0.1–4.6
	16–18 y	62	26–189	0.7–5.1	69	10–237	0.3–6.4
2	1–5 mo	*	<148	<4	*	<147	<4
	6 mo–7 y		<18	<0.5		<37	<1.0
	8–9 y		<110	<3		<110	<3
	10–12 y		<221	<6		<295	<8
	13–19 y		110–442	3–12		37–442	1–12

Male and Female

Test	Age	n	μg/dL	μmol/L
3	Premature infants	**		
	26–28 w, day 4		123–882	3.3–23.8
	31–35 w, day 4		122–710	3.3–19.2
	Full-term infants			
	3 d		88–356	2.4–9.6
	1–12 mo		a	a
	Prepubertal children			
	1–6 y		<5–57	<0.1–1.5
	6–8 y		9–72	0.2–1.9
	8–10 y		13–115	0.3–3.1

Male Age	μg/dL	μmol/L	Female[b] Age	μg/dL	μmol/L
Puberty Tanner Stage[c]			Puberty Tanner Stage[c]		
1 (<9.8 y)	13–83	0.3–2.30	1 (<9.2 y)	19–144	0.5–3.9
2 (9.8–14.5 y)	42–109	1.1–3.0	2 (9.2–13.7 y)	34–129	0.9–3.5
3 (10.7–15.4 y)	48–200	1.3–5.4	3 (10.0–14.4 y)	32–226	0.9–6.1
4 (11.8–16.2 y)	102–385	2.8–10.5	4 (10.7–15.6 y)	58–260	1.6–7.1
5 (12.8–17.3 y)	120–370	3.3–10.1	5 (11.8–18.6 y)	44–248	1.2–6.7
n**			n**		

			Male and Female		
Test	Age	n	µg/dL	µmol/L	
4	0–12 mo	90	0–1087	0.00–29.53	
	1–5 y	144	0–91	0.00–2.48	
	6–10 y	126	0–179	0.00–4.86	
	11–14 y	162	26–380	0.70–10.32	

	Male				Female		
	Age	n	µg/dL	µmol/L	n	ng/dL	nmol/L
	15–20 y	37	86–634	2.35–17.22	97	50–459	1.33–12.46

	Male				Female			
Test	Age	n	ng/dL	nmol/L	Age	n	ng/dL	nmol/L
5	0–<1 y	40	0–96	<0.003*–2.615	0–<1 y	37	0–160	<0.003*–4.360
	≥1–≤5 y	54	0–86	<0.003*–2.327	≥1–≤5 y	66	0–53	<0.003*–1.436

		Male and Female		
Age	n	µg/dL		µmol/L
>5–≤10 y	94	0–223		<0.003*–6.052

Male				Female	
Age	n	µg/dL	µmol/L	µg/dL	µmol/L
>10–≤15 y	37	8–282	0.226–7.657	—	—
>10–≤20 y	72	—	—	14–50	0.389–12.232
>15–≤20 y	20	0–503	<0.003*–13.680	—	—

Specimen Type(s)	1,4,5	Plasma/serum
	2,3	Serum
Reference(s)	1	Soldin SJ, Godwin ID, Bailey J, et al. Pediatric reference ranges for DHEA sulfate. Clin Chem 1993;39:1171. (Abstract)
	2	Babalola AA, Ellis G. Serum dehydroepiandrosterone sulfate in a normal pediatric population. Clin Biochem 1985;18:184–9.
	3	Endocrinology expected values. Esoterix Endocrinology, Calabasas Hills, CA. ©2002 Esoterix, Inc., http://www.esoterix.com
		[c]Tanner JM, Whitehouse RH. Clinical longitudinal standards for height, weight, height velocity, and the stages of puberty. Arch Dis Childhood 1976;51:170.
	4	Chan MK, Seiden-Long I, Aytekin M, et al. Canadian Laboratory Initiative on Reference Interval Database (CALIPER): pediatric reference intervals for an integrated clinical chemistry and immunoassay analyzer, Abbott ARCHITECT ci8200. Clin Biochem 2009;42:885–91.
	5	Kulasingam V, Jung BP, Blasutig IM, et al. Pediatric reference intervals for 28 chemistries and immunoassays on the Roche cobas® 6000 analyzer—A CALIPER pilot study. Clin Biochem 2010;43;1045–50.
Method(s)	1	Siemens Coat-A-Count DHEA-SO$_4$ procedure (Siemens Healthcare Diagnostics, Deerfield, IL).
	2	Used rabbit anti-DHA-3-hemisuccinyl-bovine-serum albumin antibody (Radioimmunoassay, Inc., Toronto, Canada).
	3	Not provided.
	4	Abbott Architect ci8200 (Abbott Diagnostics, Abbott Park, IL).
	5	Roche cobas® 6000 analyzer (Roche Diagnostics Limited, West Sussex,UK).
Comment(s)	1	Study used hospitalized patients and a computerized approach to removing outliers. Values are 2.5–97.5th percentiles.
	2	Study used outpatients (131 boys and 143 girls) and provides the upper reference range.
		*See reference for appropriate numbers in each age interval.
	3	**Numbers not provided.
		[a]DHEA-S levels fall to a range of 5–111 μ/dL during the first month and decrease further to a range of 5–48 μ/dL by 6 months of age.
		[b]Day of cycle not determined for prepubertal females.
		[c]As used herein, Tanner stages in males encompass development of both pubic hair and genitalia. In females, each stage encompasses both pubic hair and breast development. While this expands the chronological age range for each stage of pubertal development, it results in a better correlation with hormonal values.
	4	1459 serum/plasma specimens from children attending select outpatient clinics were collected from the five age groups noted above. Values are 2.5–97.5th percentiles.
	5	*Lower reference interval at the limit of detection of the assay.
		Approximately 600 outpatient samples from a pediatric population deemed to be metabolically stable were subdivided into five age classes ranging from 0 to 20 years of age and further partitioned by gender. Values are 2.5–97.5th percentiles.

DEOXYCORTICOSTERONE (DOC)

Age	n	Male and Female	
		ng/dL	nmol/L
Premature infants 26–28 wks, day 4	*	20–105	0.60–3.17
Newborn		a	a
Full-term infants 1–12 mo		7–49	0.21–1.48
Prepubertal children 2–10 y		2–34	0.06–1.03
Pubertal children, 8:00 a.m.		2–19	0.06–0.57

Specimen Type(s)	Serum
Reference(s)	Endocrinology expected values. Esoterix Endocrinology, Calabasas Hills, CA. ©2002 Esoterix, Inc., http://www.esoterix.com
Method(s)	Not provided.
Comment(s)	*Numbers not provided. [a]Levels are markedly elevated at birth and decrease rapidly during the first week to the range found in older infants.

11-DEOXYCORTISOL (COMPOUND S)

	Male and Female			
Test	Age	n	µg/dL	nmol/L
1	6–9 y	11	0.01–0.07	0.3–2.0
	10–11 y	17	0.01–0.09	0.3–2.6
	12–14 y	12	0.01–0.05	0.3–1.4
	15–17 y	10	0.02–0.05	0.6–1.4
	Post metyrapone reference range	*	>5.0	>144
2	Premature infants	**		
	26–28 wks, day 4		0.110–1.376	3.2–39.6
	31–35 wks, day 4		0.048–0.579	1.4–16.7
	Full-term infants			
	3 days		0.013–0.147	0.4–4.2
	1–12 mo		<0.01–0.156	0.3–4.5
	Prepubertal children, 8:00 a.m.		0.020–0.155	0.6–4.5
	Pubertal children, 8:00 a.m.		0.012–0.158	0.3–4.6

Specimen Type(s)	1,2	Serum
Reference(s)	1	Pediatric endocrine testing. Nichols Institute, 1993:14.
	2	Endocrinology expected values. Esoterix Endocrinology, Calabasas Hills, CA. ©2002 Esoterix, Inc., http://www.esoterix.com
Method(s)	1	Extraction, chromatography, radioimmunoassay.
	2	Not provided.
Comment(s)	1	Results are 2.5–97.5th percentiles.
		*Numbers not provided.
	2	**Numbers not provided.

11-DEOXYCORTISOL (COMPOUND S, METYRAPONE TEST)

		Male and Female	
Age	n	µg/dL	nmol/L
Children and adults[a]	*		
Baseline		<1	<28.9
Post metyrapone			
Single-dose test		7–18	202.3–520.2
Multiple-dose test		10–25	289.0–722.5

Specimen Type(s)	Serum
Reference(s)	Endocrinology expected values. Esoterix Endocrinology, Calabasas Hills, CA. ©2002 Esoterix, Inc., http://www.esoterix.com
Method(s)	Not provided.
Comment(s)	[a]Results in children are not significantly different than those found in adults. See Limal JM, Basmaciogullari, Rapaport R. Evaluation of single oral dose metyrapone tests in children with hypopituitarism. Acta Paed Scand 1976;65:177. *Numbers not provided.

DIHYDROTESTOSTERONE (DHT)

Age	n	Male		Female	
		ng/dL	pmol/L	ng/dL	pmol/L
Premature infants	*	10–53	345–1828	2–13	69–448
Full-term infants		5–60	172–2069	<2–15	<69–517
1–7 mo[a]		b	b	c	c
Prepubertal children		<3	<103	<3	<103

Male			Female		
Age	ng/dL	pmol/L	Age	ng/dL	pmol/L
Puberty			Puberty		
Tanner Stage[d]			Tanner Stage[d]		
1 (<9.8 y)	<3	<103	1 (<9.2 y)	<3	<103
2 (9.8–14.5 y)	3–17	103–586	2 (9.2–13.7 y)	5–12	172–414
3 (10.7–15.4 y)	8–33	276–1138	3 (10.0–14.4 y)	7–19	241–655
4 (11.8–16.2 y)	22–52	759–1793	4 (10.7–15.6 y)	4–13	138–448
5 (12.8–17.3 y)	24–65	828–2241	5 (11.8–18.6 y)	3–18	103–621
n*			n*		

Specimen Type(s)	Serum
Reference(s)	Endocrinology expected values. Esoterix Endocrinology, Calabasas Hills, CA. ©2002 Esoterix, Inc., http://www.esoterix.com [a]Data adapted from Pang S, Levine LS, Chow D, Sagiani F, Saenger P, New MI. Dihydrotestosterone and its relationship to testosterone in infancy and childhood. J Clin Endocrinol Metab 1979;48:821. [d]Tanner JM, Whitehouse RH. Clinical longitudinal standards for height, weight, height velocity, and the stages of puberty. Arch Dis Childhood 1976;51:170.
Method(s)	Not provided.
Comment(s)	*Numbers not provided. [b]DHT decreases rapidly the first week, then increases to 12–85 ng/dL between 30–60 days. Levels then decrease gradually to prepubertal values by seven months. [c]Levels decrease during the first month to <3 ng/dL and remain there until puberty. [d]As used herein, Tanner stages in males encompass development of both pubic hair and genitalia. In females, each stage encompasses both pubic hair and breast development. While this expands the chronological age range for each stage of pubertal development, it results in a better correlation with hormonal values.

DOPAMINE (URINE)

			Male and Female	
Test	Age	n	μg/g creatinine	mmol/mol creatinine
1	0–24 mo	24	<3000	<2.216
	2–4 y	37	<1533	<1.132
	5–9 y	40	<1048	<0.774
	10–19 y	41	<545	<0.403
2	<1 y	18	240–1290	0.177–0.953
	1–4 y	24	80–1220	0.059–0.901
	4–10 y	23	220–720	0.162–0.532
	10–18 y	20	120–450	0.089–0.332

Specimen Type(s)	1,2	Urine
Reference(s)	1	Soldin SJ, Lam G, Pollard A, et al. High performance liquid chromatographic analysis of urinary catecholamines employing amperometric detection: reference values and use in laboratory diagnosis of neural crest tumors. Clin Biochem 1980;13:285–91.
	2	Rosano TG. Liquid chromatographic evaluation of age related changes in the urinary excretion of free catecholamines in pediatric patients. Clin Chem 1984;30:301–3.
Method(s)	1,2	HPLC with electrochemical detection.
Comment(s)	1	Study involved healthy children. Results quoted are 95th percentile.
	2	Urine was obtained from 85 pediatric patients (diagnosis of neoplasia excluded). Results are 0–100th percentiles.

EPINEPHRINE/ADRENALINE (PLASMA)

Test	Age	n	pg/mL	nmol/L
	Male and Female			
1	2–10 d	21	36–401	0.2–2.2
	10 d–3 mo	10	55–201	0.3–1.1
	3–12 mo	14	55–438	0.3–2.4
	12–24 mo	13	36–639	0.2–3.5
	24–36 mo	8	18–438	0.1–2.4
	36 mo–15 y	20	18–456	0.1–2.5
2	30 min after birth	16	48–256	0.263–1.403
	2 h after birth	16	92–140	0.504–0.767
	3 h after birth	16	41–121	0.225–0.663
	12 h after birth	16	8–12	0.044–0.066
	24 h after birth	16	9–21	0.049–0.115
	48 h after birth	16	18–34	0.099–0.186

Specimen Type(s)	1,2	Plasma
Reference(s)	1	Candito M, Albertini M, Politano S, et al. Plasma catecholamine levels in children. J Chrom Biomed Appl 1993;617:304–7.
	2	Eliot RJ, Lam R, Leake RD, et al. Plasma catecholamine concentrations in infants at birth and during the first 48 hours of life. J Pediatrics 1980;96:311–5.
Method(s)	1	High performance liquid chromatography.
	2	Radioenzymatic method.
Comment(s)	1	Study population consisted of 86 healthy children (62 males, 24 females) aged 2 days to 15 years. Results are mean ±2 SDs.
	2	Study performed on 16 term vaginally delivered infants. Results are mean ±2 SDs.

EPINEPHRINE/ADRENALINE (URINE)

		Male and Female		
Test	Age	n	µg/g creatinine	mmol/mol creatinine
1	0–24 mo	24	<75	<0.046
	2–4 y	37	<57	<0.035
	5–9 y	40	<35	<0.022
	10–19 y	41	<34	<0.021
2	0–12 mo	18	0–375	0–0.232
	1–4 y	24	0–82	0–0.051
	4–10 y	22	5–93	0.003–0.057
	10–18 y	20	3–58	0.001–0.027

Specimen Type(s)	1,2	Urine
Reference(s)	1	Soldin SJ, Lam G, Pollard A, et al. High performance liquid chromatographic analysis of urinary catecholamines employing amperometric detection: Reference values and use in laboratory diagnosis of neural crest tumors. Clin Biochem 1980;13:285–91.
	2	Rosano TG. Liquid chromatographic evaluation of age related changes in the urinary excretion of free catecholamines in pediatric patients. Clin Chem 1984;30:301–3.
Method(s)	1,2	HPLC with electrochemical detection.
Comment(s)	1	Study involved healthy children. Results quoted are 95th percentile.
	2	Urine was obtained from 85 pediatric patients. (Patients with a discharge diagnosis of neoplasia, endocrinopathy, or muscular dystrophy were not included in the study.) Results are 0–100th percentiles.

ERYTHROPOIETIN

	Male		Female	
Age	n	mIU/mL	n	mIU/mL
1–3 y	122	1.7–17.9	97	2.1–15.9
4–6 y	89	3.5–21.9	76	2.9–8.5
7–9 y	79	1.0–13.5	80	2.1–8.2
10–12 y	98	1.0–14.0	90	1.1–9.1
13–15 y	10	2.2–14.4	148	3.8–20.5
16–18 y	66	1.5–15.2	77	2.0–14.2
Specimen Type(s)	Plasma			
Reference(s)	Krafte–Jacobs B, Williams J, Soldin SJ. Plasma erythropoietin reference ranges in children. J Pediatrics 1995;126:601–3.			
Method(s)	ELISA (Quantikine™ IVD™, Human EPO Immunoassay, R and D Systems, Minneapolis, MN).			
Comment(s)	Study used hospitalized patients and a computerized approach to removing outliers. Values are 2.5–97.5th percentiles.			

ESTRADIOL

Test	Age	Male			Female		
		n	pg/mL	pmol/L	n	pg/mL	pmol/L
1	0–30 d	38	8.0–53.0	29–194	22	12.0–66.0	44–241.6
	31–182 d	51	1.0–12.0	3.7–44	31	1.0–11.0	3.7–40.3
	6 mo–3 y	116	0.3–6.1	1.1–22.3	100	0.4–7.9	1.5–28.9
	4–6 y	43	0.9–5.5	3.3–20.1	27	1.8–8.3	6.6–30.4
	7–9 y	40	0.3–9.6	1.1–35.1	18	1.3–7.4	4.8–27.1
	10–12 y	43	0.4–10.0	1.5–36.6	27	0.3–20.6	1.1–75.4
	13–15 y	33	0.5–9.6	1.8–35.1	39	0.3–33.2	1.1–121.5
	16–18 y	14	3.6–20.4	13.2–74.7	24	4.4–49.2	16.1–180.1
2	1–6 y	*	<15	<55	*	<15	<55
	7–10 y		<15	<55		<70	<257
	11–12 y		<40	<147		10–300	37–1100
	13–15 y		<45	<165		10–300	37–1100
	16–17 y		10–50	37–184		10–300	37–1100

Test	Male				Female			
	Age	n	pg/mL	pmol/L	Age	n	pg/mL	pmol/L
3	0–<19 y	132	<20–40	<73.4–146.8	0–<6 y	50	<20–53	<73.4–194.5
					6–<11 y	103	<20–59	<73.4–216.5
					11–<15 y	61	<20–87	<73.4–319.3
					15–<19 y	55	<20–111	<73.4–407.4

Test	Male			Female		
	Age	pg/mL	pmol/L	Age	pg/mL	pmol/L
4	Newborn[a,b] (1–7 d)	c	c	Newborn[a,b] (1–7 d)	c	c
	1–6 mo[a]	d	d	1–6 mo[a]	d	d
	Prepubertal children (1–10 y)	<10	<37	Prepubertal children (1–10 y)	<10	<37
	Puberty[b]			Puberty[b,f]		
	Tanner Stage[g]			Tanner Stage[g]		
	1 (<9.8 y)	5–11	18–40	1 (<9.2 y)	5–20	18–74
	2 (9.8–14.5 y)	5–16	18–59	2 (9.2–13.7 y)	10–24	37–88
	3 (10.7–15.4 y)	5–25	18–92	3 (10.0–14.4 y)	7–60	26–220
	4 (11.8–16.2 y)	10–36	37–132	4 (10.7–15.6 y)	21–85	77–312
	5 (12.8–17.3 y)	10–36	37–132	5 (11.8–18.6 y)	34–170	125–625
	n**			n**		

Specimen Type(s)	1,3	Plasma/serum
	2	Plasma
	4	Serum
Reference(s)	1	Soldin SJ, Hicks JM, Bailey J, et al. Pediatric reference ranges for estradiol and C1 esterase inhibitor. Clin Chem 1998; 44:A17. (Abstract)
	2	Adaptation of method of Abraham GE, Odell WD, Swerdloff RS, et al. Simultaneous radioimmunoassay of plasma FSH, LH, progesterone, 17-hydroxy-progesterone and estradiol-17 beta during the menstrual cycle. J Clin Endocrinol Metab 1972;34:312–8.
		Meites S, ed. Pediatric clinical chemistry, 3rd ed. Washington, DC: AACC Press, 1989:122–3.
	3	Soldin OP, Hoffman EG, Waring MA, Soldin SJ. Pediatric reference intervals for FSH, LH, estradiol, T3, free T3, cortisol, and growth hormone on the DPC IMMULITE 1000. Clin Chim Acta 2005;355:205–10.
	4	Endocrinology expected values. Esoterix Endocrinology, Calabasas Hills, CA. ©2002 Esoterix, Inc., http://www.esoterix.com
		[a]Bidlingmaier F, Versmold H, Knorr D. Sex differences in plasma estrogen concentrations in infancy. 21 Symp Dtsch Ges Endokrin Abstract 103. Acta Endoc 1975;Suppl 193:103.
		[b]Data adapted from Bidlingmaier F, Wagner-Barnack M, Butenandt O, Knorr D. Plasma estrogens in childhood and puberty under physiologic and pathologic conditions. Pediat Res 1973;7:901.
		[g]Tanner JM, Whitehouse RH. Clinical longitudinal standards for height, weight, height velocity, and the stages of puberty. Arch Dis Childhood 1976;51:170.
Method(s)	1	Immunoassay. ACS 180 using Chiron reagents (Chiron Diagnostics Corp., East Walpole, MA).
	2	Extraction followed by RIA.
	3	Siemens IMMULITE® 1000 analyzer (Siemens Healthcare Diagnostics Deerfield, IL).
	4	Not provided.

Comment(s)	1	The study used serum/plasma from hospitalized patients and employed Chauvenet's criteria for removing outliers and a computerized approach adapted from the Hoffmann technique to obtain the 2.5–97.5th percentiles.
	2	*See pp. 122–3 of *Pediatric Clinical Chemistry* for numbers.
	3	Study used hospitalized patients and a computerized approach adapted from the Hoffmann technique. Values are the 2.5–97.5th percentiles.
	4	**Numbers not provided.
		^{c}Estradiol levels are markedly elevated at birth and fall rapidly during the first week to prepubertal values.
		^{d}Levels increase to 1.0–3.2 ng/dL between 30 and 60 days, then decline to <1.5 ng/dL by six months.
		^{e}Levels increase to 0.5–5.0 ng/dL between 30 and 60 days, then decline to 1.5 ng/dL during the first year.
		^{f}Day of cycle not determined for pubertal females.
		^{g}As used herein, Tanner stages in males encompass development of both pubic hair and genitalia. In females, each stage encompasses both pubic hair and breast development. While this expands the chronological age range for each stage of pubertal development, it results in a better correlation with hormonal values.

ESTRONE

Male			Female		
Age	ng/dL	pmol/L	Age	ng/dL	pmol/L
Newborn[a,b] (1–7 d)	c	c	Newborn[a,b] (1–7 d)	c	c
Prepubertal children (1–10 y)	<15	<56	Prepubertal children (1–10 y)	<1.5	<56
Puberty[b]			Puberty[b,d]		
Tanner Stage[e]			Tanner Stage[e]		
1 (<9.8 y)	0.5–1.7	19–63	1 (<9.2 y)	0.4–2.9	15–107
2 (9.8–14.5 y)	1.0–2.5	37–93	2 (9.2–13.7 y)	1.0–3.3	37–122
3 (10.7–15.4 y)	1.5–2.5	56–93	3 (10.0–14.4 y)	1.5–4.3	56–159
4 (11.8–16.2 y)	1.5–4.5	56–167	4 (10.7–15.6 y)	1.6–7.7	59–285
5 (12.8–17.3 y)	2.0–4.5	74–167	5 (11.8–18.6 y)	2.9–10.5	107–389
n*			n*		

Specimen Type(s)	Serum
Reference(s)	Endocrinology expected values. Esoterix Endocrinology, Calabasas Hills, CA. ©2002 Esoterix, Inc., http://www.esoterix.com
	[a]Bidlingmaier F, Versmold H, Knorr D. Sex differences in plasma estrogen concentrations in infancy. 21 Symp Dtsch Ges Endokrin Abstract 103. Acta Endoc 1975;Suppl 193:103.
	[b]Data adapted from Bidlingmaier F, Wagner-Barnack M, Butenandt O, Knorr D. Plasma estrogens in childhood and puberty under physiologic and pathologic conditions. Pediat Res 1973;7:901.
	[e]Tanner JM, Whitehouse RH. Clinical longitudinal standards for height, weight, height velocity, and the stages of puberty. Arch Dis Childhood 1976;51:170.
Method(s)	Not provided.
Comment(s)	*Numbers not provided.
	[c]Values are strikingly elevated at birth then decrease rapidly during the first week to prepubertal levels.
	[d]Day of cycle not determined for pubertal females.
	[e]As used herein, Tanner stages in males encompass development of both pubic hair and genitalia. In females, each stage encompasses both pubic hair and breast development. While this expands the chronological age range for each stage of pubertal development, it results in a better correlation with hormonal values.

		FERRITIN			
		Male		Female	
Test	Age	n	ng/mL (µg/L)	n	ng/mL (µg/L)
1	1–5 y*	44	6–24	44	6–24
	6–9 y*	50	10–55	50	10–55
	10–14 y	31	23–70	40	6–40
	14–19 y	65	23–70	110	6–40
2	1–30 d	83	6–400	66	6–515
	1–6 mo	70	6–410	50	6–340
	7–12 mo	51	6–80	51	6–45
	1–5 y	82	6–60	90	6–60
	6–19 y	77	6–320	121	6–70
3	1–30 d	83	36–381	66	36–483
	1–6 mo	70	36–391	50	36–329
	7–12 mo	51	36–100	51	36–70
	1–5 y	82	36–84	90	36–84
	6–19 y	77	36–311	121	36–92
4	0–6 wk	**	Up to 400	**	Up to 400
	7 wk–365 d		10–95		10–95
	1–9 y		10–60		10–60
	10–18 y		10–300		10–70
5	0–90 d	56	40–775	44	79–501
	91 d–12 mo	64	25–790	39	25–560
	13–36 mo	67	12–501	57	10–500
	4–10 y	77	25–280	68	22–158
	11–14 y	56	25–112	54	15–112
	15–18 y	49	18–158	49	10–125
		Male and Female			
Test	Age	n		ng/mL (µg/L)	
6	0–12 mo	94		8.7–71.6	
	1–5 y	149		3.3–127.0	
	6–10 y	127		8.8–184.7	
	11–14 y	165		5.6–216.0	
	15–20 y	151		4.4–207.0	

Specimen Type(s)	1–3,5,6	Plasma/serum
	4	Serum
Reference(s)	1	Lockitch G, Halstead A, Wadsworth L, et al. Age- and sex-specific pediatric reference intervals for zinc, copper, selenium, iron, vitamins A and E, and related proteins. Clin Chem 1988;34:1625–8.
	2	Soldin SJ, Morales A, Albalos F, et al. Pediatric reference ranges on the Abbott IMx for FSH, LH, prolactin, TSH, T_4, T_3, free T_4, Free T_3, T-uptake, and ferritin. Clin Biochem 1995;28:603–6.
	3	Murthy JN, Hicks JM, Soldin SJ. Evaluation of the Technicon Immuno I Random Access Immunoassay Analyzer and calculation of pediatric reference ranges for endocrine tests, T-uptake, and ferritin. Clin Biochem 1995;28:181–5.
	4	Dugaw KA, Jack RM, Rutledge J. Pediatric reference ranges for ferritin and AFP on the Vitros ECi analyzer. Clin Chem 2001;47:A108–9. (Abstract)
	5	Soldin OP, Bierbower LH, Choi JJ, et al. Serum iron, ferritin, transferrin, total iron binding capacity, hs-CRP, LDL cholesterol and magnesium in children; new reference intervals using the Dade Dimension Clinical Chemistry System. Clin Chim Acta 2004;342:211–7.
	6	Chan MK, Seiden-Long I, Aytekin M, et al. Canadian Laboratory Initiative on Reference Interval Database (CALIPER): pediatric reference intervals for an integrated clinical chemistry and immunoassay analyzer, Abbott ARCHITECT ci8200. Clin Biochem 2009;42:885–91.
Method(s)	1	Ferritin was measured by immunoassay and using ferrizyme reagent (Abbott Laboratories, Abbott Park, IL).
	2	IMx analyzer (Abbott Laboratories, Abbott Park, IL).
	3	Bayer Immuno I with Bayer reagents (Bayer Corp., Tarrytown NY).
	4	Chemiluminescent immunoassay, Vitros ECi (Ortho-Clinical Diagnostics, Raritan, NJ).
	5	The ferritin (FERR) method for the Dimension® RxL clinical chemistry system is a one-step enzyme immunoassay based on the "sandwich" principle. Siemens Dimension RxL Analyzer (Siemens Healthcare Diagnostics, Deerfield, IL).
	6	Abbott Architect ci8200 (Abbott Diagnostics, Abbott Park, IL).
Comment(s)	1	The study population was healthy children. Non-parametric methods were used to determine the 0.025 and 0.975 fractiles.
		*No significant differences were found for males and females. These ranges were therefore derived from combined data.
	2,3	Study used hospitalized patients and a computerized approach adapted from the Hoffmann technique. Values are 2.5–97.5th percentiles.
	4	**Ages ranged from 1 h to 18 y with a total of 92 specimens.
	5	Study used hospitalized patients and a computerized approach adapted from the Hoffmann technique to obtain the 2.5–97.5th percentiles.
	6	1459 serum/plasma specimens from children attending select outpatient clinics were collected from the five age groups noted above. Values are 2.5–97.5th percentiles.

FOLIC ACID

Test	Age	Male n	Male nmol/L	Female n	Female nmol/L
1	0–1 y	111	16.3–50.8	73	14.3–51.5
	2–3 y	105	5.7–34.0	135	3.9–35.6
	4–6 y	154	1.1–29.4	104	6.1–31.9
	7–9 y	103	5.2–27.0	102	5.4–30.4
	10–12 y	105	3.4–24.5	90	2.3–23.1
	13–18 y	127	2.7–19.9	159	2.7–16.3
2	Newborn[a]	*	16–72	*	16–72
	After newborn period		4–20		4–20
	Adult[a,b,c]		10–63		10–63

Specimen Type(s)	1	Plasma
	2	Serum
Reference(s)	1	Hicks JM, Cook J, Godwin ID, et al. Vitamin B12 and folate: pediatric reference ranges. Arch Pathol Lab Med, 1993;117:704–6.
	2	[a]Behrman RE, Vaughan VC, eds. Nelson textbook of pediatrics. Philadelphia, PA: WB Saunders Company, 1983:1827–60.
		[b]Nathan DG, Oski FA. Hematology of infancy and childhood, 4th ed. Appendix IX. Philadelphia, PA: WB Saunders Company, 1993.
		[c]Hall CA, Bardwell SA, Allen ES, et al. Variation in plasma folate levels among groups of healthy persons. Am J Clin Nutr 1975;28:854–7.
Method(s)	1	Radioimmunoassay Quantaphase (BioRad, Hercules, CA).
	2	See reference.
Comment(s)	1	Study used hospitalized patients and a computerized approach adapted from the Hoffmann technique to obtain the 2.5–97.5th percentiles.
	2	*See references for particulars of assays used and numbers studied.

FOLLICLE-STIMULATING HORMONE (FSH)

Test	Age	Male		Female	
		n	U/L (mIU/mL)	n	U/L (mIU/mL)
1	<2 y	110	0.2–1.8	33	0.2–6.6
	2–5 y	124	0.2–1.4	96	0.2–3.8
	6–10 y	99	0.2–1.3	155	0.2–2.7
	11–20 y	100	0.2–8.0	247	0.2–8.0
2	<2 y	110	0.4–2.1	33	0.4–7.1
	2–5 y	124	0.4–1.7	94	0.4–4.2
	6–10 y	99	0.4–1.6	136	0.4–3.0
	11–20 y	100	0.4–8.7	263	0.4–8.6

Test	Male			Female		
	Age	n	U/L (mIU/mL)	Age	n	U/L (mIU/mL)
3	13–<19 y	62	<0.1–8.6	0–<6 y	51	<0.1–7.1
				6–<11 y	106	<0.1–4.3
				11–<15 y	86	<0.1–12.0
				15–<19 y	112	<0.1–11.0
4	1–9 y	*	0.0–5.0	1–2 y	*	0.0–8.0
	10–11 y		0.0–6.0	3–8 y		0.0–5.0
	13–18 y		0.0–10.0	9–11 y		0.0–10.0
				12–18 y		0.0–15.0
5	0–<1 y	40	<0.1*–3.25	0–<1 y	37	1.17–12.51
	≥1–≤5 y	58	<0.1*–1.86	≥1–≤10 y	133	0.46–5.98
	>5–≤10 y	49	<0.1*–2.32	>10–≤15 y	63	0.87–8.86
	>10–≤15 y	63	0.57–6.89			

Male and Female		
Age	n	U/L
>15–≤20 y	62	0.70–9.69

Specimen Type(s)	1–3,5	Plasma/serum
	4	Serum
Reference(s)	1	Soldin SJ, Morales A, Albalos F, et al. Pediatric reference ranges on the Abbott IMx for FSH, LH, prolactin, TSH, T_4, T_3, free T_4, free T_3, T-uptake, IgE and ferritin. Clin Biochem 1995;28:603–6.
	2	Murthy JN, Hicks JM, Soldin SJ. Evaluation of the Technicon Immuno I Random Access Immunoassay Analyzer and calculation of pediatric reference ranges for endocrine tests, T-uptake, and ferritin. Clin Biochem 1995;28:181–5.
	3	Soldin OP, Hoffman EG, Waring MA, Soldin SJ. Pediatric reference intervals for FSH, LH, estradiol, T_3, free T_3, cortisol, and growth hormone on the DPC IMMULITE 1000. Clin Chim Acta 2005;355:205–10.
	4	Dugaw KA, Jack RM, Rutledge J. Pediatric reference ranges for FSH and LH on the Vitros ECi analyzer. Clin Chem 2001;47:A108. (Abstract)
	5	Kulasingam V, Jung BP, Blasutig IM, et al. Pediatric reference intervals for 28 chemistries and immunoassays on the Roche cobas® 6000 analyzer—A CALIPER pilot study. Clin Biochem 2010;43;1045–50.
Method(s)	1	IMx (Abbott Laboratories, Abbott Park, IL).
	2	Technicon Immuno I with Bayer reagents (Bayer Corp., Tarrytown, NY).
	3	Siemens IMMULITE® 1000 analyzer (Siemens Healthcare Diagnostics, Deerfield, IL).
	4	Chemiluminescent immunoassay, Vitros ECi (Ortho-Clinical Diagnostics, Raritan, NJ).
	5	Roche cobas® 6000 analyzer (Roche Diagnostics Limited, West Sussex, UK).
Comment(s)	1, 2	Study used hospitalized patients and a computerized approach to removing outliers. Values are 2.5–97.5th percentiles.
	3	Study used hospitalized patients and a computerized approach adapted from the Hoffmann technique. Values are the 2.5–97.5th percentiles.
	4	*Ages ranged from 1 h to 18 y with a total of 99 specimens.
	5	*Lower reference interval at the limit of detection of the assay. Approximately 600 outpatient samples from a pediatric population deemed to be metabolically stable were subdivided into five age classes ranging from 0 to 20 years of age and further partitioned by gender. Values are 2.5–97.5th percentiles.

FREE FATTY ACIDS

Male and Female		
Age	n	mmol/L
1–12 mo	12	0.5–1.6
1–7 y	27	0.6–1.5
7–15 y	9	0.2–1.1

Specimen Type(s)	Plasma
Reference(s)	Bonnefont JP, Specola NB, Vassault A, et al. The fasting test in pediatrics: application to the diagnosis of pathological hypo- and hyperketotic states. Eur J Pediatr 1990;150:80–5.
Method(s)	Standard enzymatic procedure. See reference.
Comment(s)	Results are 10–90th percentiles.

FRUCTOSAMINE

Test	Age	n	mmol/L
	Male and Female		
1	0–3 y	33	1.56–2.27
	3–6 y	49	1.73–2.34
	6–9 y	35	1.82–2.56
	9–12 y	24	2.04–2.50
	12–15 y	29	2.02–2.63
2	Normal	203	0.174–0.286
	Controlled diabetics	80	0.210–0.421
	Uncontrolled diabetics	123	0.268–0.870

Specimen Type(s)	1, 2	Serum
Reference(s)	1	De Schepper J, Derde MP, Goubert P, et al. Reference values for fructosamine concentrations in children's sera: influence of protein concentration, age and sex. Clin Chem 1988;34:2444–7.
	2	RoTAG™ Fructosamine (glycated protein) Assay from package insert, Roche diagnostics kit.
Method(s)	1	Roche diagnostics kit on Cobas-Bio (Roche Diagnostic Systems, Inc., Branchburg, NJ).
	2	Roche diagnostics kit (Roche Diagnostic Systems, Inc., Branchburg, NJ).
Comment(s)	1	Results are 2.5–97.5th percentiles.
	2	Results using the new Roche kit using polylysine differ by a factor of approximately 10 from the old kit using DMF equivalents. Ages not provided in study.

GAMMA-GLUTAMYLTRANSFERASE (GGT)

Test	Age	Male n	Male U/L	Female n	Female U/L
1	1–182 d	109	12–122	67	15–132
	183–365 d	36	1–39	36	1–39
	1–12 y	488	3–22	391	4–22
	13–18 y	170	2–42	208	4–24
2*	1–7 d	137	25–148	102	19–131
	8–30 d	186	23–153	129	17–124
	1–3 mo	172	17–130	189	17–124
	4–6 mo	241	8–83	191	15–109
	7–12 mo	199	10–35	190	10–54
3*	1–3 y	50**	5–16	50**	5–16
	4–6 y	40**	8–18	40**	8–18
	7–9 y	80**	11–21	80**	11–21
	10–11 y	27	14–25	34	14–23
	12–13 y	31	14–37	49	12–21
	14–15 y	26	10–28	52	12–22
	16–19 y	40	9–29	61	9–23
4	1–7 d	***	25–168	***	18–148
	8–30 d		23–174		16–140
	1–3 mo		16–147		16–140
	4–6 mo		5–93		13–123
	7–12 mo		8–38		8–59
	1–3 y		2–15		2–15
	4–6 y		5–17		5–17
	7–9 y		9–20		9–20
	10–11 y		12–25		12–23
	12–13 y		12–39		10–20
	14–15 y		8–29		10–22
	16–19 y		6–30		6–23

Specimen Type(s)	1,2,4	Plasma/serum
	3	Plasma
Reference(s)	1	Soldin SJ, Hicks JM, Bailey J, et al. Pediatric reference ranges for gammaglutamyltransferase. Clin Chem 1997;43:S198. (Abstract)
	2	Soldin SJ, Savwoir TV, Guo Y. Pediatric reference ranges for gamma-glutamyltransferase and urea nitrogen during the first year of life on the Vitros 500 analyzer. Clin Chem 1997;43:S199. (Abstract)
	3	Lockitch G, Halstead AC, Albersheim S, et al. Age- and sex-specific pediatric reference intervals for biochemistry analytes as measured with the Ektachem-700 analyzer. Clin Chem 1988;34:1622–5.
	4	Ghoshal AK, Soldin SJ. Evaluation of the Dade Behring Dimension RxL: integrated chemistry system-pediatric reference ranges. Clin Chim Acta 2003;331:135–46.
Method(s)	1	Substrate used is L-α-glutamyl-3-carboxy-4-nitroanilide and rate of production of 5-amino-2-nitrobenzoate is measured on the Hitachi 747. Boehringer Mannheim reagents (Boehringer Mannheim Diagnostics, Indianapolis, IN).
	2,3	Vitros 500 (2) and 700 (3) (Ortho-Clinical Diagnostics, Raritan, NJ). α-Glutamyl-p-nitroanilide used as substrate.
	4	Gamma-glutamyl transferase catalyzes the transfer of the glutamyl moiety from gamma-glutamyl-3-carboxy-4-nitranilide to glycylglycine thereby releasing 5-amino-2-nitrobenzoate, which absorbs at 405 nm. Siemens Dimension RxL Analyzer (Siemens Healthcare Diagnostics, Deerfield, IL).
Comment(s)	1,2	Study used hospitalized patients and a computerized approach adapted from the Hoffmann technique. Values are 2.5–97.5th percentiles.
	3	The study population was healthy children. Non-parametric methods were used to determine the 0.025 and 0.975 fractiles. *The values originally published were multiplied by 0.8383 to accommodate changes made by the manufacturer that decreased values by ~16%. **Males and females were not studied separately.
	4	***Reference ranges were obtained by comparing results from previously published data and using regression equations.

GASTRIN

Male and Female		
Age	n	ng/L pg/mL
Newborns (1–12 d)	*	69–190
Infants (1.5–22 mo)		55–186
Prepubertal and pubertal children, fasting		
3–4 h		2–168
5–6 h		3–117
≥8 h		1–125

Specimen Type(s)	Serum
Reference(s)	Pediatric endocrine testing. Nichols Institute, 1993:20.
Method(s)	Radioimmunoassay.
Comment(s)	Results are 2.5–97.5th percentiles. *See reference for numbers.

GHRELIN

Male and Female		
Age	n	ng/L
28–36 wk gestation	62	695–1157
37–41 wk gestation	15	356–1294

Specimen Type(s)	Serum
Reference(s)	Siahanidou T, Mandyla H, Vounatsou M, Anagnostakis D, Papassotiriou I, Chrousos GP. Circulating peptide YY concentrations are higher in preterm than full-term infants and correlate negatively with body weight and positively with serum ghrelin concentrations. Clin Chem 2005;51:2131–7.
Method(s)	Serum ghrelin was measured using a commercial RIA (Phoenix Pharmaceuticals, Inc., Burlingame, CA).
Comment(s)	Values are 2.5–97.5th percentiles.

GLOBULINS TOTAL, CALCULATED

Test	Age	Male			Female		
		n	g/dL	g/L	n	g/dL	g/L
1	1–182 d	58	1.3–2.4	13–24	51	1.3–2.1	13–24
	183–365 d	29	1.7–3.0	17–30	42	1.2–2.4	12–24
	1–3 y	134	2.1–3.4	21–34	186	1.8–3.3	18–33
	4–6 y	139	2.2–3.4	22–34	114	2.0–3.6	20–36
	7–9 y	118	2.2–3.6	22–36	107	2.2–3.4	22–34
	10–12 y	9	2.4–3.4	24–34	82	2.3–3.3	23–33
	13–15 y	7	2.1–3.7	21–37	83	2.4–3.8	24–38
	16–18 y	5	2.0–3.9	20–39	54	2.6–3.6	26–36
2	<1 y	104*	0.4–3.7	4–37	104*	0.4–3.7	4–37
	1–3 y	247*	1.6–3.5	16–35	247*	1.6–3.5	16–35
	4–9 y	694*	1.9–3.4	19–34	694*	1.9–3.4	19–34
	10–49 y	17905*	1.9–3.5	19–35	17905*	1.8–3.5	18–35

Specimen Type(s)	1	Plasma
	2	Serum
Reference(s)	1	Hicks JM, Bjorn S, Beatey J, et al. Pediatric reference ranges for albumin, globulin and total protein on the Hitachi 747. Clin Chem 1995;41:S93. (Abstract)
	2	Appleton C. Queensland Medical Laboratory, Brisbane, Australia. Meites, S, ed. Pediatric clinical chemistry, 3rd ed. Washington, DC: AACC Press, 1989:132.
Method(s)	1	Calculated (total protein–albumin).
	2	Calculated. Obtained by subtracting albumin values from total protein values obtained on the Technicon SMAC II Analyzer (Technicon Instruments Corp., Tarrytown, NY)..
Comment(s)	1	Study used hospitalized patients and a computerized approach adapted from the Hoffmann technique. Values are 2.5–97.5th percentiles.
	2	Values are 2.5–97.5th percentiles. *Numbers listed include males and females.

GLUCAGON

	Male and Female	
Age	n	ng/L pg/mL
Cord blood	6	0–215
3 d	12	0–1750
4–14 y	9	0–148

Specimen Type(s)	Plasma
Reference(s)	Pediatric endocrine testing. Nichols Institute, 1993:20.
Method(s)	Extraction, radioimmunoassay.
Comment(s)	Results are 2.5–97.5th percentiles.

GLUCOSE

		Male and Female			
Test	Age	n	mg/dL	mmol/L	
1	0–1 mo	207	55–115	3.1–6.4	
	1–6 mo	96	57–117	3.2–6.5	
2	Outside the neonatal period	482*	70–126	3.9–7.0	

			Male		Female	
Test	Age	n	mg/dL	mmol/L	mg/dL	mmol/L
3	0–1 d	**	36–110	2.00–6.11	36–89	2.00–4.94
	1–7 d		47–110	2.61–6.11	47–110	2.61–6.11
	>7 d		54–117	3.00–6.49	54–117	3.00–6.49

Specimen Type(s)	1	Whole blood
	2	Serum
	3	Plasma/serum
Reference(s)	1	Snell J, Greeley C, Colaco A, et al. Pediatric reference ranges for arterial pH whole blood electrolytes and glucose. Clin Chem 1993;39:1173. (Abstract)
	2	Lockitch G, Halstead AC, Albersheim S, et al. Age- and sex-specific pediatric reference intervals for biochemistry analyses as measured with the Ektachem-700 analyzer. Clin Chem 1988;34:1622–5.
	3	Ghoshal AK, Soldin SJ. Evaluation of the Dade Behring Dimension RxL: integrated chemistry system-pediatric reference ranges. Clin Chim Acta 2003;331:135–46.
Method(s)	1	Glucose oxidase YSI 2300 (Yellow Springs Instruments, Yellow Springs, OH).
	2	Glucose oxidase, Vitros 700 (Ortho-Clinical Diagnostics, Raritan, NJ).
	3	The glucose method is an adaptation of the hexokinase-glucose-6-phosphate dehydrogenase method. Siemens Dimension RxL Analyzer (Siemens Healthcare Diagnostics, Deerfield, IL).
Comment(s)	1	Results are 2.5–97.5th percentiles. The 0- to 1-mo group includes an appreciable number of premature infants.
	2	Results are 2.5–97.5th percentiles.
		*No significant differences were found for males and females. These ranges were therefore derived from combined data.
	3	**Numbers not provided.
		Reference ranges were obtained by comparing results from previously published data and using regression equations.

GLUCOSE (URINE)

		Male and Female	
Age	n	mg/dL	mmol/L
All ages	*	6–19	0.33–1.05

Specimen Type(s)	Urine
Reference(s)	Ghoshal AK, Soldin SJ. Evaluation of the Dade Behring Dimension RxL: integrated chemistry system-pediatric reference ranges. Clin Chim Acta 2003;331:135–46.
Method(s)	The glucose method is an adaptation of the hexokinase-glucose-6-phosphate dehydrogenase method. Siemens Dimension RxL Analyzer (Siemens Healthcare Diagnostics, Deerfield, IL).
Comment(s)	*Numbers not provided. Reference ranges were obtained by comparing results from previously published data and using regression equations.

GLUCOSE CEREBROSPINAL FLUID (CSF GLUCOSE)

Test	Age	n	Male and Female	
1	All ages	*	60–80% of blood glucose	
Test	Age	n	mg/dL	mmol/L
2	All ages	**	41–84	2.28–4.66

Specimen Type(s)	1,2	Cerebrospinal fluid
Reference(s)	1	Meites S, ed. Pediatric clinical chemistry, 2nd ed. Washington, DC: AACC Press, 1981:216–8.
	2	Ghoshal AK, Soldin SJ. Evaluation of the Dade Behring Dimension RxL: integrated chemistry system-pediatric reference ranges. Clin Chim Acta 2003;331:135–46.
Method(s)	2	The glucose method is an adaptation of the hexokinase-glucose-6-phosphate dehydrogenase method. Siemens Dimension RxL Analyzer (Siemens Healthcare Diagnostics, Deerfield, IL).
Comment(s)	1	Glucose is measured in CSF primarily in the diagnosis of bacterial meningitis. CSF glucose should be compared to blood glucose for interpretation. *Numbers not provided.
	2	**Numbers not provided. Reference ranges were obtained by comparing results from previously published data and using regression equations.

GLUTATHIONE PEROXIDASE ACTIVITY

	Male and Female	
Age	n	U/L
Newborn (term)	*	180–890
1–5 y		554–985
6–15 y		567–1153
16 y–adult		780–1269
Specimen Type(s)	Serum	
Reference(s)	Lockitch G. Trace elements in pediatrics. JIFCC 1996:9;46–51.	
Method(s)	Enzymatic.	
Comment(s)	*Numbers of individuals included in each age range not provided. In selenium deficiency, activity of glutathione peroxidase in plasma/serum reflects selenium status.	

\multicolumn{4}{c}{**GROWTH HORMONE**}			
\multicolumn{4}{c}{Male and Female}			
Test	Age	n	µg/L
1	After stimulation by exercise, arginine, or insulin	*	>5
2	0–<7 y	56	<1–13.6
	7–<11 y	99	<1–16.4
	11–<15 y	155	<1–14.4
	15–<19 y	90	<1–13.4
Specimen Type(s)	1	Serum	
	2	Plasma/serum	
Reference(s)	1	Reference values and SI unit information. The Hospital for Sick Children, Toronto, Canada, 1993:369.	
	2	Soldin OP, Hoffman EG, Waring MA, Soldin SJ. Pediatric reference intervals for FSH, LH, estradiol, T3, free T3, cortisol, and growth hormone on the DPC IMMULITE 1000. Clin Chim Acta 2005;355:205–10.	
Method(s)	1	RIA.	
	2	Siemens IMMULITE® 1000 analyzer (Siemens Healthcare Diagnostics, Deerfield, IL).	
Comment(s)	1	*Numbers not provided. Differential diagnosis of short stature, slow growth, or evaluation of pituitary function. Random samples have little diagnostic value.	
	2	Study used hospitalized patients and a computerized approach adapted from the Hoffmann technique. Values are the 2.5–97.5th percentiles.	

GROWTH HORMONE (URINE)

Test	Age	Male			Female		
		n	ng/24h	ng/g creatinine	n	ng/24h	ng/g creatinine
1	Tanner 1, 2.2–13.3 y	*	0.4–6.3	0.9–12.3	*	0.4–6.3	0.9–12.3
	Tanner 2, 10.3–14.6 y		0.8–12.0	1.0–14.1		0.8–12.0	1.0–14.1
	Tanner 3, 11.5–15.3 y		1.7–20.4	1.9–17.0		1.7–20.4	1.9–17.0
	Tanner 4, 12.7–17.1 y		1.5–18.2	1.3–14.4		1.5–18.2	1.3–14.4
	Tanner 5, 13.5–19.9 y		1.2–14.5	0.8–11.0		1.2–14.5	0.8–11.0
	Adults		0.6–20.9	0.4–15.9		0.6–20.9	0.4–15.9
2	3–4 y	**		>6.5	**		>9.1
	5–6 y			>8.7			>8.9
	7–8 y			>6.8			>5.9
	9–10 y			>5.6			>8.9
	11–12 y			>6.2			>9.3
	13–14 y			>10.5			>6.3
	15–16 y			>3.8			>5.1
	17–18 y			>3.1			>3.1
	19–20 y			>1.9			>1.9

Specimen Type(s)	1, 2	Urine
Reference(s)	1	Main K, Philips M, Jergensen M, et al. Urinary growth hormone excretion in 657 healthy children and adults. Normal values, inter- and intraindividual variations. Horm Res 1991;36:174–82.
	2	Nukada O, Moriwake T, Kanzaki S, et al. Age-related changes in urinary growth hormone level and its clinical application. Acta Paediatr Jpn 1990;32:32–8.
Method(s)	1	ELISA, antipituitary hGH (Nanormon® and recombinant hGH Norditropin®) (Novo Nordisk A/S, Gentofte, Denmark).
	2	EIA (Sumitomo Chemical Co., Ltd., Takarazuka, Japan).
Comment(s)	1	547 healthy children and 110 healthy adults. *See reference for numbers. Results are 2.5–97.5th percentiles.
	2	Study evaluated urinary GH secretion in 270 normal subjects aged 3–20 years. Results provided are the lower limit of normal (Mean–2 SDs). **See reference for numbers.

HAPTOGLOBIN

	Male and Female		
Age	n	mg/dL	µmol/L
0–12 mo	92	0–222	0–22.2
1–5 y	146	0–243	0–24.3
6–10 y	126	0–208	0–20.8
11–14 y	159	0–226	0–22.6
15–20 y	147	0–236	0–23.6
Specimen Type(s)	Plasma/serum		
Reference(s)	Chan MK, Seiden-Long I, Aytekin M, et al. Canadian Laboratory Initiative on Reference Interval Database (CALIPER): pediatric reference intervals for an integrated clinical chemistry and immunoassay analyzer, Abbott ARCHITECT ci8200. Clin Biochem 2009;42:885–91		
Method(s)	Abbott Architect ci8200 (Abbott Diagnostics, Abbott Park, IL).		
Comment(s)	1459 serum/plasma specimens from children attending select outpatient clinics were collected from the five age groups noted above. Values are 2.5–97.5th percentiles.		

HEMOGLOBIN A$_{1c}$ (HbA$_{1c}$)

		Male and Female		
Test	Age		n	%
1	Normal individuals		*	4–7
	Pregnant women			5–8
	Old age			6–9
	Stable diabetics			8–10
	Young or unstable diabetics			8–18
2	Normal individuals		**	3.4–6.1
3	Healthy non-diabetics		***	4–6
	Diabetics with average control			9–10
4	Normal		****	3.7–6.2

Specimen Type(s)	1–3	Whole blood
	4	Plasma/serum
Reference(s)	1	Kellen JA. Disorders of carbohydrate metabolism. In: AG Gornall, ed. Applied biochemistry of clinical disorders, 2nd ed. Philadelphia, PA: JB Lippincott Company, 1986:379–402.
	2	Children's National Medical Center, Washington, DC. Unpublished data.
	3	Reference values and SI unit information. The Hospital for Sick Children, Toronto, Canada, 1993.
	4	Ghoshal AK, Soldin SJ. Evaluation of the Dade Behring Dimension RxL: integrated chemistry system-pediatric reference ranges. Clin Chim Acta 2003;331:135–46.
Method(s)	1	Not provided.
	2	Immunochemical technique DCA 2000 Hemoglobin A1C System. (Miles Diagnostics Division, Elkhart, IN).
	3	High-performance liquid chromatography.
	4	Uses Siemens reagents. Siemens Dimension RxL Analyzer (Siemens Healthcare Diagnostics, Deerfield, IL).
Comment(s)	1	*Numbers not provided.
	2	**Numbers not provided. Hemoglobin is glycated in proportion to the average blood glucose concentration over the life span of the red cell (90–120 d). Hemoglobin A$_{1c}$ allows tracking of metabolic control in individual patients over time.
	3	***Numbers not provided.
	4	****Numbers not provided. Reference ranges were obtained by comparing results from previously published data and using regression equations.

HIGH-DENSITY LIPOPROTEIN CHOLESTEROL (HDL-C)

Test	Age	Male n	Male mg/dL	Male mmol/L	Female n	Female mg/dL	Female mmol/L
1	1–9 y	50*	35–82	0.91–2.12	50*	35–82	0.91–2.12
	10–13 y	56*	36–84	0.93–2.17	56*	36–84	0.93–2.17
	14–19 y	60*	35–65	0.91–1.68	60*	35–65	0.91–1.68
2	5 y	**	37–72	0.96–1.86	**	34–69	0.88–1.78
	10 y		31–70	0.80–1.81		34–69	0.88–1.78
	15 y		29–61	0.75–1.58		34–69	0.88–1.78
	20 y		29–58	0.75–1.50		36–70	0.93–1.81
3	0–24 mo	43***	8–61	0.21–1.58	43***	8–61	0.21–1.58
	2–<7 y	95	23–70	0.60–1.81	71	12–64	0.31–1.66
	7–<12 y	99	25–79	0.65–2.05	133	23–80	0.60–2.07
	12–<16 y	199	19–76	0.49–1.97	181	25–83	0.65–2.15
	16–<19 y	91	26–75	0.67–1.94	157	21–77	0.54–1.99
4	0–24 mo	****	12–60	0.31–1.55	****	12–60	0.31–1.55
	2–<7 y		26–68	0.67–1.76		16–62	0.41–1.61
	7–<12 y		28–76	0.73–1.97		26–77	0.67–1.99
	12–<16 y		22–73	0.57–1.89		28–79	0.73–2.05
	16–<19 y		28–72	0.73–1.86		24–74	0.62–1.92
5	0–12 mo	56	14–63	0.36–1.62	72	19–69	0.49–1.78

	Male and Female			
	Age	n	mg/dL	mmol/L
	1–5 y	150	25–73	0.64–1.90
	6–10 y	131	32–74	0.82–1.91
	11–14 y	154	28–72	0.73–1.87

		Male			Female		
	Age	n	mg/dL	mmol/L	n	mg/dL	mmol/L
	15–20 y	36	25–76	0.65–1.98	145	34–75	0.87–1.93

Test	Male and Female			
6	Age	n	mg/dL	mmol/L
	0–≤15 y	499	26–75	0.67–1.94

		Male			Female		
	Age	n	mg/dL	mmol/L	n	mg/dL	mmol/L
	>15–≤20 y	34	25–68	0.66–1.76	49	25–89	0.64–2.16

Specimen Type(s)	1	Serum
	2	Plasma
	3–6	Plasma/serum
Reference(s)	1	Lockitch G, Halstead AC, Albersheim S, et al. Age- and sex-specific pediatric reference intervals for biochemistry analytes as measured with the Ektachem-700 analyzer. Clin Chem 1988;34:1622–5.
	2	Kottke BA, Moll PP, Michels VV, et al. Levels of lipids, lipoproteins and apolipoproteins in a defined population. Mayo Clin Proc 1991;66:1198–208.
	3	Murthy JN, Soldin SJ. Pediatric reference ranges for HDL-cholesterol on the Vitros 500 analyzer. Clin Chem 1999;45;A23. (Abstract)
	4	Ghoshal AK, Soldin SJ. Evaluation of the Dade Behring Dimension RxL: integrated chemistry system-pediatric reference ranges. Clin Chim Acta 2003;331:135–46.
	5	Chan MK, Seiden-Long I, Aytekin M, et al. Canadian Laboratory Initiative on Reference Interval Database (CALIPER): pediatric reference intervals for an integrated clinical chemistry and immunoassay analyzer, Abbott ARCHITECT ci8200. Clin Biochem 2009;42:885–91.
	6	Kulasingam V, Jung BP, Blasutig IM, et al. Pediatric reference intervals for 28 chemistries and immunoassays on the Roche cobas® 6000 analyzer—A CALIPER pilot study. Clin Biochem 2010;43;1045–50.
Method(s)	1,3	Dextran sulfate precipitation/cholesterol oxidase. Vitros 700 (1) and 500 (3) (Ortho-Clinical Diagnostics, Raritan, NJ).
	2	LDL and VLDL precipitated using polyethylene glycol 6000. Supernatant used to determine HDL-cholesterol by an enzymatic method. See reference.
	4	The HDL cholesterol assay is a homogeneous method for directly measuring HDL levels without the need for off-line pretreatment or centrifugation steps. Siemens Dimension RxL Analyzer (Siemens Healthcare Diagnostics, Deerfield, IL).
	5	Abbott Architect ci8200 (Abbott Diagnostics, Abbott Park, IL).
	6	Roche cobas® 6000 analyzer (Roche Diagnostics Limited, West Sussex, UK).
Comment(s)	1	Healthy children. Results are 2.5–97.5th percentiles. *No significant differences were found for males and females. These ranges were therefore derived from combined data.
	2	Results are 5–95th percentiles read off graphs in reference. **See reference for numbers.
	3	***Male and female combined. The study used hospitalized patients and employed Chauvenet's criteria for removing outliers and a computerized approach adapted from the Hoffmann technique to obtain the 2.5–97.5th percentiles. The 2.5th and 97.5th percentiles are lower during the first 2 y of life. For children >2 y, the 2.5th percentiles are lower than those previously described by Lockitch, et al. Clin Chem 1988;34:1622–5.
	4	****Reference ranges were obtained by comparing results from previously published data and using regression equations.
	5	1459 serum/plasma specimens from children attending select outpatient clinics were collected from the five age groups noted above. Values are 2.5–97.5th percentiles.
	6	Approximately 600 outpatient samples from a pediatric population deemed to be metabolically stable were subdivided into five age classes ranging from 0 to 20 years of age and further partitioned by gender. Values are 2.5–97.5th percentiles.

HOMOCYSTEINE (TOTAL)

Test	Age	Male and Female	
		n	µmol/L
1	2 mo–10 y	105	3.3–8.3
	11–15 y	59	4.7–10.3
	16–18 y	31	4.7–11.3
2	0–<1 mo	66	3.0–8.5

	Male		Female	
Age	n	µmol/L	n	µmol/L
1–<6 mo	—	—	59	4.4–15.1
1 mo–<1 y	88	3.5–8.5	—	—
6 mo–<1 y	—	—	61	3.0–13.2
1–<3 y	49	2.6–11.0	68	3.6–10.7
3–<10 y	103	3.6–11.0	71	3.0–9.8
10–<13 y	54	4.3–9.3	37	3.6–8.9

Male and Female		
Age	n	mmol/L
13–<15 y	76	4.7–10.6

	Male		Female	
Age	n	µmol/L	n	µmol/L
15–<18 y	60	4.7–14.5	91	4.7–13.2
>18 y	41	5.1–19.0	—	—

Specimen Type(s)	1	Plasma
	2	Plasma/serum
Reference(s)	1	Vilaseca MA, Moyano D, Ferrer I, et al. Total homocysteine in pediatric patients. Clin Chem 1997;43:690–2.
	2	Soldin OP, Dahlin JR, Gresham EG, et al. IMMULITE 2000 age and sex-specific reference intervals for alpha fetoprotein, homocysteine, insulin, insulin-like growth factor-1, insulin-like growth factor binding protein-3, C-peptide, immunoglobulin E and intact parathyroid hormone. Clin Biochem 2008;41:937–42.
Method(s)	1	HPLC
	2	IMMULITE® 2000 immunoassay system. (Siemens Healthcare Diagnostics, Deerfield, IL).
Comment(s)	1	This study was performed on plasma obtained from healthy children. Results are the 2.5th and 97.5th percentiles.
	2	Study was conducted at both Children's National Medical Center and Georgetown University, Washington, DC. Results were obtained from the Children's National Medical Center laboratory information system over the period January 5, 2001, to March 8, 2007. Patient results were accessed and used to establish reference intervals. All patient identifiers were removed except age and sex. Data was analyzed using the Hoffmann approach, and was computer adapted. Values are 2.5–97.5th percentiles.

HOMOVANILLIC ACID (HVA; 4-HYDROXY-3-METHOXYPHENYLACETIC ACID) (URINE)

Test	Age	n	mg/g creatinine	mmol/mol creatinine	n	mg/24h	µmol/24h
1	0–1 y	37	<32.6	<20.2	48	<2.8	<15.4
	2–4 y	49	<22.0	<13.6	34	<4.7	<25.8
	5–9 y	79	<15.1	<9.4	20	<5.4	<29.6
	10–19 y	55	<12.8	<7.9	40	<7.2	<39.5
	>19 y	56	<7.6	<4.7	56	<8.3	<45.6
2	0–3 mo	12	11.3–35.0	7.0–21.7			
	3–12 mo	30	8.4–44.9	5.2–27.8			
	1–2 y	16	12.2–31.8	7.6–19.7			
	2–5 y	21	3.4–32.0	2.1–19.8			
	5–10 y	25	6.8–23.7	4.2–14.7			
	10–15 y	13	3.2–13.6	2.0–8.4			
	>15 y	9	3.2–9.6	2.0–6.0			

(Male and Female)

Specimen Type(s)	1,2	Urine
Reference(s)	1	Soldin SJ, Hill JG. Liquid chromatographic analysis for urinary 4-hydroxy-3-methoxy-mandelic acid and 4-hydroxy-3-methoxy-phenylacetic acid and its use in investigation of neural crest tumors. Clin Chem 1981;27:502–3.
	2	Tuchman M, Morris CL, Ramnaraine ML, et al. Value of random urinary homovanillic acid and vanillyl mandelic acid levels in the diagnosis and management of patients with neuroblastoma: comparison with 24-h urine collections. Pediatrics 1985;75:324–8.
Method(s)	1	HPLC with electrochemical detection.
	2	Capillary gas chromatography.
Comment(s)	1	Analysis performed on patients under investigation of hypertension or not suspected of having a neural crest tumor. All patients studied free of neoplasia. Results are 95th percentile.
	2	Values are 0–100th percentiles. Normal values were obtained from 93 pediatric patients in whom the diagnosis of neural crest tumors had been excluded.

HYALURONIC ACID

Male and Female

Age	n	µg/L
1–3 mo	*	49–153
2–3 y		9–40
4–18 y		6–32

Specimen Type(s)	Serum
Reference(s)	Trivedi P, Cheeseman P, Mowat AP. Serum hyaluronic acid in healthy infants and children and its value as a marker of progressive hepatobiliary disease starting in infancy. Clin Chim Acta 1993;215:29–39.
Method(s)	Affinity-binding radiometric assay (Pharmacia Diagnostics AB, Uppsala, Sweden).
Comment(s)	397 healthy infants. *See reference for details.

β–HYDROXYBUTYRATE/ 3–HYDROXYBUTYRATE

Male and Female

Age	n	mg/dL	mmol/L
1–12 mo	12	1.0–10.0	0.1–1.0
1–7 y	27	<1.0–9.4	<0.1–0.9
7–15 y	9	<1.0–3.1	<0.1–0.3

Specimen Type(s)	Whole blood precipitated with perchloric acid.
Reference(s)	Bonnefont JP, Specola NB, Vassault A, et al. The fasting test in paediatrics: application to the diagnosis of pathological hypo- and hyperketotic states. Eur J Pediatr 1990;150:80–5.
Method(s)	Standard enzymatic procedure. See reference.
Comment(s)	The above results are 10–90th percentiles and refer to 15-h fasting values. For 20- and 24-h fasting values see reference.

5–HYDROXYINDOLEACETIC ACID (5HIAA) (URINE)

	Male and Female			
Age	n	mg/24h	μmol/24h	mg/g creatinine
3–8 y	76	0.4–5.6	2.1–29.3	1.2–16.2
9–12 y	43	1.0–6.3	5.2–32.9	2.4–8.7
13–17 y	43	0.9–6.5	4.7–34.0	1.8–5.5
Adults	51	1.0–7.0	5.2–36.6	1.3–6.9

Specimen Type(s)	Urine
Reference(s)	Pediatric endocrine testing. Nichols Institute, 1993:24.
Method(s)	Fluorescent polarization.
Comment(s)	Results are 2.5–97.5th percentiles.

17α-HYDROXYPROGESTERONE (17α-OHP)

Test	Age	Male			Female		
		n	ng/dL	nmol/L	n	ng/dL	nmol/L
1	1–30 d	45	53–186	1.6–5.6	51	17–204	0.5–6.2
	31–182 d	48	35–157	1.1–4.8	46	25–110	0.8–3.3
	183–365 d	27	6–40	0.2–1.2	34	5–47	0.2–1.4
	1–3 y	104	2–19	0.1–0.6	77	3–51	0.1–1.5
	4–6 y	91	1–34	0.0–1.0	64	4–34	0.1–1.0
	7–9 y	74	1–45	0.0–1.4	53	4–44	0.1–1.3
	10–12 y	89	1–34	0.0–1.0	45	3–33	0.1–1.0
	13–15 y	61	23–82	0.7–2.5	42	2–72	0.1–2.2
	16–18 y	33	8–100	0.2–3.0	38	3–91	0.1–2.8

Test	Age	Male and Female		
		n	ng/dL	nmol/L
2	Premature infants	*		
	26–28 wks, day 4		124–841	3.7–25.2
	31–35 wks, day 4		26–568	0.8–17.0
	Full-term infants	*		
	3 d		7–77	0.2–2.3
	1–12 mo		13–106[a]	0.4–3.23[a]
	Prepubertal children			
	1–10 y	*	3–90	0.1–2.7[b]

Male				Female[b]			
Age	n	ng/dL	nmol/L	Age	n	ng/dL	nmol/L
Puberty[c]	*			Puberty[c]	*		
Tanner Stage[d]				Tanner Stage[d]			
1 (<9.8 y)		3–90	0.1–2.7	1 (<9.2 y)		3–82	0.1–2.5
2 (9.8–14.5 y)		5–115	0.2–3.5	2 (9.2–13.7 y)		11–98	0.3–2.9
3 (10.7–15.4 y)		10–138	0.3–4.1	3 (10.0–14.4 y)		11–155	0.3–4.7
4 (11.8–16.2 y)		29–180	0.9–5.4	4 (10.7–15.6 y)		18–230	0.5–6.9
5 (12.8–17.3 y)		24–175	0.7–5.3	5 (11.8–18.6 y)		20–265	0.6–8.0

Test	Male and Female			
	Age	n	ng/dL	nmol/L
3	0–<6 mo	119	25–248	0.8–7.5

Male				Female			
Age	n	ng/dL	nmol/L	Age	n	ng/dL	nmol/L
6 mo–<18 y	153	7–100	0.2–3.0	6 mo–<6 y	132	3–107	0.1–3.2
				6–<10 y	182	6–62	0.2–1.9
				10–<18 y	144	15–137	0.5–4.1

Specimen Type(s)	1	Plasma/serum
	2, 3	Serum
Reference(s)	1	Soldin SJ, Bailey J, Beatey J, et al. Pediatric reference ranges for 17α-hydroxyprogesterone. Clin Chem 1995;41:S92. (Abstract)
	2	Endocrinology expected values. Esoterix Endocrinology, Calabasas Hills, CA. ©2002 Esoterix, Inc., http://www.esoterix.com
		[c]Data adapted from Bidlingmaier F, Wagner-Barnack M, Butenandt O, Knorr D. Plasma estrogrens in childhood and puberty under physiologic and pathologic conditions. Pediat Res 1973;7:901
		[d]Tanner JM, Whitehouse RH. Clinical longitudinal standards for height, weight, height velocity, and the stages of puberty. Arch Dis Childhood 1976;51:170.
	3	Soldin OP, Sharma H, Husted L, Soldin SJ. Pediatric reference intervals for aldosterone, 17alpha-hydroxyprogesterone, dehydroepoandrosterone, testosterone and 25-hydroxy vitamin D3 using tandem mass spectrometry. Clin Biochem 2009;42:823–7.
Method(s)	1	Coat-A-count (Diagnostic Products Corporation, Los Angeles, CA).
	2	Not provided.
	3	Samples were analyzed using isotope dilution liquid chromatography tandem mass spectrometry (LC/MS/MS.
Comment(s)	1	Study used hospitalized patients and a computerized approach adapted from the Hoffmann technique. Values are 2.5–97.5th percentiles.
	2	*Numbers not provided.
		[a]Levels increase after the first week to peak values ranging from 40–200 ng/dL between 30 and 60 days. Values then decline to prepubertal range before one year.
		[b]Days of cycle not determined for prepubertal females.
		[d]As used herein, Tanner stages in males encompass development of both pubic hair and genitalia. In females, each stage encompasses both pubic hair and breast development. While this expands the chronological age range for each stage of pubertal development, it results in a better correlation with hormonal values.
	3	Reference intervals were determined for neonates and children 0–18 years of age. The study was conducted using outpatient samples obtained between January 1, 2004, and June 30, 2008. Values are the 2.5–97.5th percentiles.

IMMUNOGLOBULIN A (IgA)

Test	Age	Male and Female		
		n	mg/dL	g/L
1	0–12 mo	75	0–83	0.00–0.83
	1–3 y	52	20–100	0.20–1.00
	4–6 y	41	27–195	0.27–1.95
	7–9 y	55	34–305	0.34–3.05
	10–11 y	38	53–204	0.53–2.04
	12–13 y	38	58–359	0.58–3.59
	14–15 y	38	47–249	0.47–2.49
	16–19 y	74	61–348	0.61–3.48

Test	Age	Male			Female		
		n	mg/dL	g/L	n	mg/dL	g/L
2	1–30 d	61	1–17	0.01–0.17	56	1–16	0.01–0.16
	31–182 d	52	6–47	0.06–0.47	60	1–49	0.01–0.49
	183–365 d	37	8–89	0.08–0.89	23	13–75	0.13–0.75
	1–3 y	105	15–142	0.15–1.42	127	21–117	0.21–1.17
	4–6 y	138	50–192	0.50–1.92	135	39–170	0.39–1.70
	7–9 y	111	64–209	0.64–2.09	105	34–181	0.34–1.81
	10–12 y	104	51–223	0.51–2.23	115	61–198	0.61–1.98
	13–15 y	104	35–252	0.35–2.52	118	68–246	0.68–2.46
	16–18 y	112	74–261	0.74–2.61	100	75–267	0.75–2.67
3	1–30 d	*	0–11	0–0.11	*	0–10	0–0.10
	31–182 d		0–40	0–0.40		0–42	0–0.42
	183–365 d		1–82	0.01–0.82		6–68	0.06–0.68
	1–3 y		9–137	0.09–1.37		15–111	0.15–1.11
	4–6 y		44–187	0.44–1.87		33–166	0.33–1.66
	7–9 y		58–204	0.58–2.04		28–180	0.28–1.80
	10–12 y		46–218	0.46–2.18		55–193	0.55–1.93
	13–15 y		29–251	0.29–2.51		62–241	0.62–2.41
	16–18 y		68–259	0.68–2.59		69–262	0.69–2.62

Specimen Type(s)	1	Plasma
	2,3	Plasma/serum
Reference(s)	\multicolumn{2}{l	}{The above values have been adjusted from those published in references 1 and 2 to convert results to current IFCC units using IFCC guidelines/standards.}
	1	Lockitch G, Halstead AC, Quigley G, et al. Age- and sex-specific pediatric reference intervals: Study design and methods illustrated by measurement of serum proteins with the Behring LN nephelometer. Clin Chem 1988;34:1618–21.
	2	Soldin SJ, Bailey J, Beatey J, et al. Pediatric reference ranges for immunoglobulins G, A and M on the Behring nephelometer. Clin Chem 1996;42:S308. (Abstract)
	3	Ghoshal AK, Soldin SJ. Evaluation of the Dade Behring Dimension RxL: integrated chemistry system-pediatric reference ranges. Clin Chim Acta 2003;331:135–46.
Method(s)	1,2	Nephelometry (Behring Diagnostics, Inc., Westwood, MA).
	3	The IgA method is a quantitative turbidimetric assay using endpoint detection based on the precipitation of IgA by its polyclonal antibody. Siemens Dimension RxL Analyzer (Siemens Healthcare Diagnostics, Deerfield, IL).
Comment(s)	1	Study performed on healthy normal children. Results are 2.5–97.5th percentiles.
	2	Study used hospitalized patients and a computerized approach adapted from the Hoffmann technique to obtain the 2.5–97.5th percentiles.
	3	*Numbers not provided.
		Reference ranges were obtained by comparing results from previously published data and using regression equations.

IMMUNOGLOBULIN D (IgD)

Male and Female			
Age	n	mg/dL	mg/L
Newborn	*	None detected	None detected
Thereafter		0–8	0–80

Specimen Type(s)	Serum
Reference(s)	Behrman RE, ed. Nelson textbook of pediatrics, 14th ed. Philadelphia, PA: WB Saunders Company, 1992:1813.
Method(s)	Not given.
Comment(s)	*Numbers not provided in reference.

IMMUNOGLOBULIN E (IgE)

Test	Age	Male and Female		
		n	KIU/L	
1*	0–<1 y	50	<0.1**–8	

Test	Age	Male		Female	
		n	KIU/L	n	KIU/L
	≥1–≤5 y	50	0.8–542	72	1.3–260
	>5–≤15 y	104	3–1555	120	0.7–509
	>15–≤20 y	12	<0.1**–41	30	1.5–197
2	0–12 mo	40	2–24	29	0–20
	1–3 y	74	2–149	104	2–55
	4–10 y	155	4–249	124	8–279
	11–15 y	137	7–280	152	5–295
	16–18 y	58	5–268	67	7–698
3	0–12 mo	28	<12	28	<8
	1–3 y	44	<90	67	<28
	4–10 y	76	<163	56	<137
	11–18 y	96	<179	109	<398
4	0–12 mo	***	<20	***	17
	1–3 y		<85		33
	4–10 y		<146		124
	11–18 y		<159		177
5	0–<4 mo	63	1–40	31	1–24
	4 mo–<1 y	125	1–126	98	2–63
	1–<3 y	356	3–398	250	3–141
	3–<10 y	556	4–999	396	3–398
	10–<13 y	169	8–631	152	9–407
	13–<15 y	94	10–562	112	23–355
	15–18 y	108	11–447	181	5–355

Specimen Type(s)	1–5	Plasma/serum
Reference(s)	1	Kulasingam V, Jung BP, Blasutig IM, et al. Pediatric reference intervals for 28 chemistries and immunoassays on the Roche cobas® 6000 analyzer—A CALIPER pilot study. Clin Biochem 2010;43;1045–50.
	2	Soldin SJ, Morales A, Albalos F, et al. Pediatric reference ranges on the Abbott IMx for FSH, LH, prolactin, TSH, T_4, T_3, free T_4, free T_3, T-uptake, IgE and ferritin. Clin Biochem 1995;28:603–6.
	3	Soldin SJ, Lenherr S, Kumar A. Pediatric reference ranges for IgE. Clin Chem 1995;41:S92. (Abstract)
	4	DPC IMMULITE® 1000 reference ranges used at Children's National Medical Center, Washington, DC.
	5	Soldin OP, Dahlin JR, Gresham EG, et al. IMMULITE 2000 age and sex-specific reference intervals for alpha fetoprotein, homocysteine, insulin, insulin-like growth factor-1, insulin-like growth factor binding protein-3, C-peptide, immunoglobulin E and intact parathyroid hormone. Clin Biochem 2008;41:937–42.
Method(s)	1	Roche cobas® 6000 analyzer (Roche Diagnostics Limited, West Sussex, UK).
	2	Abbott IMx (Abbott Laboratories, Abbott Park, IL).
	3	Behring LN nephelometer (Behring Diagnostics, Westwood, MA).
	4	Siemens IMMULITE® 1000 analyzer (Siemens Healthcare Diagnostics, Deerfield, IL).
	5	IMMULITE® 2000 immunoassay system. (Siemens Healthcare Diagnostics, Deerfield, IL).
Comment(s)	1	*IgE reference intervals derived assuming a healthy population with no significant allergic reactions. **Lower reference interval at the limit of detection of the assay. Approximately 600 outpatient samples from a pediatric population deemed to be metabolically stable were subdivided into five age classes ranging from 0 to 20 years of age and further partitioned by gender. Values are 2.5–97.5th percentiles.
	2,3	Study performed on hospitalized patients and the Hoffmann technique applied to obtain the 2.5–97.5th percentiles (1) and 97.5th percentile (2).
	4	*Numbers not provided. Pediatric reference ranges were determined by using regression equations to transform the previously established ranges on the Bayer Immuno 1 (Bayer Corp., Tarrytown, NY) based on the Hoffmann technique to their corresponding values on the IMMULITE® 1000.
	5	Study was conducted at both Children's National Medical Center and Georgetown University, Washington, DC. Results were obtained from the Children's National Medical Center laboratory information system over the period January 5, 2001, to March 8, 2007. Patient results were accessed and used to establish reference intervals. All patient identifiers were removed except age and sex. Data was analyzed using the Hoffmann approach, and was computer adapted. Values are 2.5–97.5th percentiles.

IMMUNOGLOBULIN G (IgG)

Test	Age	Male and Female		
		n	mg/dL	g/L
1	0–12 mo	75	232–1411	2.32–14.11
	1–3 y	52	453–916	4.53–9.16
	4–6 y	41	504–1465	5.04–14.65
	7–9 y	55	572–1474	5.72–14.74
	10–11 y	38	698–1560	6.98–15.60
	12–13 y	38	759–1550	7.59–15.50
	14–15 y	38	716–1711	7.16–17.11
	16–19 y	74	549–1584	5.49–15.84

Test	Age	Male			Female		
		n	mg/dL	g/L	n	mg/dL	g/L
2	1–30 d	61	221–838	2.21–8.38	56	188–876	1.88–8.76
	31–182 d	52	166–547	1.66–5.47	60	332–675	3.32–6.75
	183–365 d	37	156–829	1.56–8.29	23	346–658	3.46–6.58
	1–3 y	105	431–1109	4.31–11.09	127	468–1196	4.68–11.96
	4–6 y	138	485–1318	4.85–13.18	135	574–1309	5.74–13.09
	7–9 y	111	595–1428	5.95–14.28	105	501–1459	5.01–14.59
	10–12 y	104	695–1602	6.95–16.02	115	599–1590	5.99–15.90
	13–15 y	104	603–1582	6.03–15.82	118	757–1621	7.57–16.21
	16–18 y	112	537–1682	5.37–16.82	100	810–1792	8.10–17.92
3	1–30 d	*	197–833	1.97–8.33	*	162–872	1.62–8.72
	31–182 d		140–533	1.40–5.33		311–664	3.11–6.64
	183–365 d		130–823	1.30–8.23		325–647	3.25–6.47
	1–3 y		413–1112	4.13–11.12		451–1202	4.51–12.02
	4–6 y		468–1328	4.68–13.28		560–1319	5.60–13.19
	7–9 y		582–1441	5.82–14.41		485–1473	4.85–14.73
	10–12 y		685–1620	6.85–16.20		586–1609	5.86–16.09
	13–15 y		590–1600	5.90–16.00		749–1640	7.49–16.40
	16–18 y		522–1703	5.22–17.03		804–1817	8.04–18.17

Specimen Type(s)	1	Plasma
	2,3	Plasma/serum
Reference(s)		The above values have been adjusted from those published in references 1 and 2 to convert results to current IFCC units using IFCC guidelines/standards.
	1	Lockitch G, Halstead AC, Quigley G, et al. Age- and sex-specific pediatric reference intervals: Study design and methods illustrated by measurement of serum proteins with the Behring LN nephelometer. Clin Chem 1988;34:1618–21.
	2	Soldin SJ, Bailey J, Beatey J, et al. Pediatric reference ranges for immunoglobulins G, A and M on the Behring nephelometer. Clin Chem 1996;42:S308. (Abstract)
	3	Ghoshal AK, Soldin SJ. Evaluation of the Dade Behring Dimension RxL: integrated chemistry system-pediatric reference ranges. Clin Chim Acta 2003;331:135–46.
Method(s)	1,2	Nephelometry (Behring Diagnostics, Inc., Westwood, MA).
	3	The IgG method is a quantitative turbidimetric assay using endpoint detection based on the precipitation of IgG by its polyclonal antibody. Siemens Dimension RxL Analyzer (Siemens Healthcare Diagnostics, Deerfield, IL).
Comment(s)	1	Study performed on healthy normal children. Results are 2.5–97.5th percentiles.
	2	Study used hospitalized patients and a computerized approach adapted from the Hoffmann technique to obtain the 2.5–97.5th percentiles.
	3	*Numbers not provided. Reference ranges were obtained by comparing results from previously published data and using regression equations.

IMMUNOGLOBULIN G SUBCLASS (IgG-SUBCLASS)

Male and Female, g/L (To convert to mg/dL, multiply by 100)					
Age	n	IgG_1	IgG_2	IgG_3	IgG_4
Cord blood serum preterm	20	3.4–9.7	0.7–1.7	0.2–0.5	0.2–0.7
Cord blood serum term	20	5.8–13.7	0.6–5.2	0.2–1.2	0.2–1.0
5 y	20	5.6–12.7	0.4–4.4	0.3–1.0	0.1–0.8
6 y	20	6.2–11.3	0.5–4.0	0.3–0.8	0.2–0.9
7 y	20	5.4–10.5	0.9–3.5	0.3–1.1	0.2–1.1
8 y	20	5.6–10.5	0.7–4.5	0.2–1.1	0.1–0.8
9 y	20	3.9–11.4	0.7–4.7	0.4–1.2	0.2–1.0
10 y	20	4.4–10.8	0.6–4.0	0.3–1.2	0.1–0.9
11 y	20	6.4–10.9	0.9–4.3	0.3–0.9	0.2–1.0
12 y	20	6.0–11.5	0.9–4.8	0.4–1.0	0.2–0.9
13 y	20	6.1–11.5	0.9–7.9	0.2–1.1	0.1–0.8
Specimen Type(s)	Serum				
Reference(s)	Miles J, Riches P. The determination of IgG subclass concentrations in serum by enzyme linked immunosorbent assay: establishment of age-related reference ranges for cord blood samples, children aged 5–13 years and adults. Ann Clin Biochem 1994;31:24–8.				
Method(s)	ELISA.				
Comment(s)	Cord blood samples were obtained from both premature and term infants. Samples were obtained from school children and adults. Results are 5–95th percentiles.				

IMMUNOGLOBULIN LIGHT CHAINS (KAPPA AND LAMBDA)

		Male and Female			
		Kappa		Lambda	
Age	n	mg/dL	g/L	mg/dL	g/L
Newborn	325	770–870	7.7–8.7	170–190	1.7–1.9
Premature	168	310–490	3.1–4.9	160–190	1.6–1.9
1 mo	50	360–480	3.6–4.8	160–200	1.6–2.0
2 mo	50	240–270	2.4–2.7	160–170	1.6–1.7
3 mo	50	170–250	1.7–2.5	100–130	1.0–1.3
4 mo	50	190–240	1.9–2.4	80–100	0.8–1.0
5 mo	50	200–320	2.0–3.2	100–110	1.0–1.1
1 y	50	300–530	3.0–5.3	90–120	0.9–1.2
2 y	50	360–640	3.6–6.4	130–150	1.3–1.5
3 y	50	410–540	4.1–5.4	110–140	1.1–1.4
4 y	50	480–830	4.8–8.3	120–170	1.2–1.7
5 y	50	540–710	5.4–7.1	130–170	1.3–1.7
6 y	50	600–850	6.0–8.5	160–170	1.6–1.7
7 y	50	460–750	4.6–7.5	140–180	1.4–1.8
8 y	50	730–910	7.3–9.1	140–170	1.4–1.7
9 y	50	710–960	7.1–9.6	150–180	1.5–1.8
10 y	50	820–960	8.2–9.6	130–190	1.3–1.9
11 y	50	600–830	6.0–8.3	140–180	1.4–1.8
12 y	50	820–960	8.2–9.6	170–200	1.7–2.0
13 y	50	820–1050	8.2–10.5	140–170	1.4–1.7
14 y	50	800–1080	8.0–10.8	190–220	1.9–2.2
15 y	50	720–1100	7.2–11.0	170–230	1.7–2.3
16 y	50	700–1120	7.0–11.2	150–220	1.5–2.2

Specimen Type(s)	Serum
Reference(s)	Herkner KR, Salzer H, Böck A, et al. Pediatric and perinatal reference intervals for immunoglobulin light chains kappa and lambda. Clin Chem 1992;38:548–50.
Method(s)	Nephelometry. Kallestad Model CPM 300 (Diagnostics Pasteur, Marnes-la-Coquette, France).
Comment(s)	The subjects were healthy newborns and children who had been screened to rule out immunological disorders and infections. The ranges are 10–90th percentiles.

IMMUNOGLOBULIN M (IgM)

Test	Age	Male and Female		
		n	mg/dL	g/L
1	0–12 mo	75	0–145	0.00–1.45
	1–3 y	52	19–146	0.19–1.46
	4–6 y	41	24–210	0.24–2.10
	7–9 y	55	32–208	0.32–2.08
	10–11 y	38	31–180	0.31–1.80
	12–13 y	38	35–239	0.35–2.39
	14–15 y	38	15–188	0.15–1.88
	16–19 y	74	23–259	0.23–2.59

Test	Age	Male			Female		
		n	mg/dL	g/L	n	mg/dL	g/L
2	1–30 d	61	8–78	0.08–0.78	56	13–70	0.13–0.70
	31–182 d	52	18–98	0.18–0.98	60	6–142	0.06–1.42
	183–365 d	37	27–132	0.27–1.32	23	3–145	0.03–1.45
	1–3 y	105	42–161	0.42–1.61	127	47–200	0.47–2.00
	4–6 y	138	43–166	0.43–1.66	135	54–200	0.54–2.00
	7–9 y	111	33–155	0.33–1.55	105	42–181	0.42–1.81
	10–12 y	104	39–167	0.39–1.67	115	54–228	0.54–2.28
	13–15 y	104	38–200	0.38–2.00	118	46–242	0.46–2.42
	16–18 y	112	40–195	0.40–1.95	100	58–241	0.58–2.41
3	1–30 d	*	0–65	0–0.65	*	1–57	0–0.57
	31–182 d		6–84	0.06–0.84		0–127	0–1.27
	183–365 d		15–117	0.15–1.17		0–130	0–1.30
	1–3 y		30–146	0.30–1.46		35–184	0.35–1.84
	4–6 y		31–151	0.31–1.51		42–184	0.42–1.84
	7–9 y		21–140	0.21–1.40		30–165	0.30–1.65
	10–12 y		27–151	0.27–1.51		42–211	0.42–2.11
	13–15 y		26–184	0.26–1.84		34–225	0.34–2.25
	16–18 y		28–179	0.28–1.79		45–224	0.45–2.24

Specimen Type(s)	1	Plasma
	2,3	Plasma/serum
Reference(s)	\multicolumn{2}{l}{The above values have been adjusted from those published in references 1 and 2 to convert results to current IFCC units using IFCC guidelines/standards.}
	1	Lockitch G, Halstead AC, Quigley G, et al. Age- and sex-specific pediatric reference intervals: Study design and methods illustrated by measurement of serum proteins with the Behring LN nephelometer. Clin Chem 1988;34:1618–21.
	2	Soldin SJ, Bailey J, Beatey J, et al. Pediatric reference ranges for immunoglobulins G, A and M on the Behring nephelometer. Clin Chem 1996;42:S308. (Abstract)
	3	Ghoshal AK, Soldin SJ. Evaluation of the Dade Behring Dimension RxL: integrated chemistry system-pediatric reference ranges. Clin Chim Acta 2003;331:135–46.
Method(s)	1,2	Nephelometry (Behring Diagnostics, Inc., Westwood, MA).
	3	The IgM method is a quantitative turbidimetric assay using endpoint detection based on the precipitation of IgM by its polyclonal antibody. Siemens Dimension RxL Analyzer (Siemens Healthcare Diagnostics, Deerfield, IL).
Comment(s)	1	Study performed on healthy normal children. Results are 2.5–97.5th percentiles.
	2	Study used hospitalized patients and a computerized approach adapted from the Hoffmann technique to obtain the 2.5–97.5th percentiles.
	3	*Numbers not provided.
		Reference ranges were obtained by comparing results from previously published data and using regression equations.

INSULIN

Test	Age	n	pmol/L		μIU/mL	
	Male and Female					
1	0–12 mo	90	9.0–199.6		1.3–28.7	
	1–5 y	145	4.8–205.6		0.7–29.6	
	6–10 y	127	11.9–400.9		1.7–57.7	
	11–14 y	162	18.3–754.3		2.6–108.6	
	15–20 y	139	14.2–725.9		2.0–104.5	
2	0–<1 y	35	16.0–191.0		2.3–27.5	
	1–<3 y	34	11.1–191.0		1.6–27.5	

		Male			**Female**		
Test	Age	n	pmol/L	μIU/mL	n	pmol/L	μIU/mL
	3–<10 y	297	20.1–214.6	2.9–30.9	323	27.8–282.7	4.0–40.7
	10–<13 y	465	43.8–347.9	6.3–50.1	465	54.9–390.3	7.9–56.2
	13–<15 y	288	54.9–347.9	7.9–50.1	402	54.9–390.3	7.9–56.2
	15–18 y	334	49.3–459.1	7.1–66.1	527	51.4–356.3	7.4–51.3
3	0–<1 y	36	5.2–169.1	0.8–24.3	54	5.7–258.7	0.8–37.2

		Male and Female				
	Age	n	pmol/L		μIU/mL	
	≥1–≤5 y	28	1.8–226.4		0.3–32.6	

		Male			**Female**		
	Age	n	pmol/L	μIU/mL	n	pmol/L	μIU/mL
	>5–≤20 y	135	10.0–326.6	1.4–47.0	191	9.7–678.7	1.4–97.7

Specimen Type(s)	1–3	Plasma/serum
Reference(s)	1	Chan MK, Seiden-Long I, Aytekin M, et al. Canadian Laboratory Initiative on Reference Interval Database (CALIPER): pediatric reference intervals for an integrated clinical chemistry and immunoassay analyzer, Abbott ARCHITECT ci8200. Clin Biochem 2009;42:885–91.
	2	Soldin OP, Dahlin JR, Gresham EG, et al. IMMULITE 2000 age and sex-specific reference intervals for alpha fetoprotein, homocysteine, insulin, insulin-like growth factor-1, insulin-like growth factor binding protein-3, C-peptide, immunoglobulin E and intact parathyroid hormone. Clin Biochem 2008;41:937–42.
	3	Kulasingam V, Jung BP, Blasutig IM, et al. Pediatric reference intervals for 28 chemistries and immunoassays on the Roche cobas® 6000 analyzer—A CALIPER pilot study. Clin Biochem 2010;43;1045–50.
Method(s)	1	Abbott Architect ci8200 (Abbott Diagnostics, Abbott Park, IL).
	2	IMMULITE® 2000 immunoassay system. (Siemens Healthcare Diagnostics, Deerfield, IL).
	3	Roche cobas® 6000 analyzer (Roche Diagnostics Limited, West Sussex, UK)
Comment(s)	1	1459 serum/plasma specimens from children attending select outpatient clinics were collected from the five age groups noted above. Values are 2.5–97.5th percentiles.
	2	Study was conducted at both Children's National Medical Center and Georgetown University, Washington, DC. Results were obtained from the Children's National Medical Center laboratory information system over the period January 5, 2001, to March 8, 2007. Patient results were accessed and used to establish reference intervals. All patient identifiers were removed except age and sex. Data was analyzed using the Hoffmann approach, and was computer adapted. Values are 2.5–97.5th percentiles.
	3	Approximately 600 outpatient samples from a pediatric population deemed to be metabolically stable were subdivided into five age classes ranging from 0 to 20 years of age and further partitioned by gender. Values are 2.5–97.5th percentiles.

INSULIN-LIKE GROWTH FACTOR BINDING PROTEIN-3 (IGF BINDING PROTEIN-3, IGF BP-3)

Test	Age	Male and Female	
		n	mg/L (µg/mL)
1	2–23 mo	40	0.7–2.3
	2–7 y	36	0.9–4.1
	8–11 y	68	1.5–4.3
	12–18 y	207	2.2–4.2
	19–55 y	137	2.0–4.0
2	0–1 wk	32	0.42–1.39
	1–4 wk	16	0.77–2.09
	1–3 mo	20	0.87–2.54
	3–6 mo	18	0.98–2.64
	6–12 mo	25	1.07–2.76
	1–3 y	36	1.41–2.97
	3–5 y	22	1.52–3.32
	5–7 y	25	1.66–3.59
	7–9 y	47	1.82–3.80
	9–11 y	50	2.12–4.26
	11–13 y	77	2.22–4.89
	13–15 y	77	2.31–5.24
	15–17 y	42	2.33–4.95
	20–30 y	32	2.20–4.93

Test	Age	Male		Female	
		n	mg/L (µg/mL)	n	mg/L (µg/mL)
3	1–<3 y	60	1.3–3.5	60	1.3–3.5
	3–<10 y	117	2.0–5.0	102	2.9–5.3
	10–<13 y	68	3.0–5.8	49	3.2–6.0
	13–<15 y	80	3.6–6.0	—	—
	13–15 y	—	—	35	3.3–6.3
	>15 y	—	—	46	3.3–6.6
	15–18 y	52	3.5–6.8	—	—
	>18 y	42	3.3–6.6	—	—

Specimen Type(s)	1,2	Serum
	3	Plasma/serum
Reference(s)	1	Pediatric endocrine testing. Nichols Institute, 1993;26.
	2	Blum WF, Ranke MB, Kietzmann K, et al. A specific radio-immunoassay for the growth hormone (GH)-dependent somatomedin-binding protein: Its use for diagnosis of GH deficiency. J Clin Endocrinol Metab 1990;70:1292–8.
	3	Soldin OP, Dahlin JR, Gresham EG, et al. IMMULITE 2000 age and sex-specific reference intervals for alpha fetoprotein, homocysteine, insulin, insulin-like growth factor-1, insulin-like growth factor binding protein-3, C-peptide, immunoglobulin E and intact parathyroid hormone. Clin Biochem 2008;41:937–42.
Method(s)	1,2	Radioimmunoassay.
	3	IMMULITE® 2000 immunoassay system (Siemens Healthcare Diagnostics, Deerfield, IL).
Comment(s)	1	Results are 2.5–97.5th percentiles.
	2	Results are 5–95th percentiles.
	3	Study was conducted at both Children's National Medical Center and Georgetown University, Washington, DC. Results were obtained from the Children's National Medical Center laboratory information system over the period January 5, 2001, to March 8, 2007. Patient results were accessed and used to establish reference intervals. All patient identifiers were removed except age and sex. Data was analyzed using the Hoffmann approach, and was computer adapted. Values are 2.5–97.5th percentiles.

INSULIN-LIKE GROWTH FACTOR-I (IGF-I), SOMATOMEDIN-C

Test	Age	Male			Female		
		n	nmol/L	ng/mL	n	nmol/L	ng/mL
1	1–30 d	34	0.2–7.3	2–56	31	0.9–11.9	7–92
	31–182 d	55	0.2–10.6	2–82	61	0.6–9.4	5–72
	183–365 d	27	0.2–7.0	2–54	35	1.1–9.8	8–75
	1–3 y	140	2.2–15.0	17–116	111	1.6–17.5	12–135
	4–6 y	126	2.7–20.8	21–160	83	1.7–22.8	13–176
	7–9 y	100	8.4–26.9	65–207	74	6.9–32.9	53–253
	10–12 y	113	9.0–25.4	69–196	86	9.8–46.3	75–357
	13–15 y	116	10.4–52.5	80–404	94	8.6–44.7	66–344
	16–18 y	62	12.5–49.7	96–383	85	12.8–47.3	99–364
2	2 mo–5 y	97	2.2–32.2	17–248	97	2.2–32.2	17–248
	6–8 y	40	11.4–61.5	88–474	40	11.4–61.5	88–474
	9–11 y	125	14.2–73.4	110–565	174	15.2–100.1	117–771
	12–15 y	135	26.2–124.3	202–957	195	33.9–142.3	261–1096
	16–24 y	95	23.6–101.3	182–780	95	23.6–101.3	182–780
3	3 mo–<1 y	30	3.9–12.8	30–98	30	3.9–12.8	30–98
	1–<3 y	45	4.2–11.9	32–91	42	5.8–22.7	44–174
	3–<10 y	227	5.9–38.6	45–295	200	8.5–59.7	65–457
	10–<13 y	143	13.1–57.1	100–437	127	13.1–90.5	100–692
	13–<15 y	136	17.3–90.5	132–692	87	19.3–119.2	148–912
	15–18 y	126	16.9–96.9	129–741	58	21.2–92.6	162–708
	>18 y	60	11.4–73.5	87–562	41	9.0–64.1	69–490

Specimen Type(s)	1,2	Serum
	3	Plasma/serum
Reference(s)	1	Soldin SJ, Hicks JM, Bailey J, et al. Pediatric reference ranges for creatine kinase and insulin-like growth factor 1. Clin Chem 1997;43:S199. (Abstract)
	2	Nichols Institute Test Catalogue, 1996.
	3	Soldin OP, Dahlin JR, Gresham EG, et al. IMMULITE 2000 age and sex-specific reference intervals for alpha fetoprotein, homocysteine, insulin, insulin-like growth factor-1, insulin-like growth factor binding protein-3, C-peptide, immunoglobulin E and intact parathyroid hormone. Clin Biochem 2008;41:937–42.
Method(s)	1	Radioimmunoassay, IncStar® (IncStar Corporation, Stillwater, MN).
	2	Radioimmunoassay.
	3	IMMULITE® 2000 immunoassay system. (Siemens Healthcare Diagnostics, Deerfield, IL).
Comment(s)	1	Results are 2.5–95th percentiles. Study used hospitalized patients and a computerized adaptation of the Hoffmann technique.
	2	Results are 2.5–97.5th percentiles.
	3	Study was conducted at both Children's National Medical Center and Georgetown University, Washington, DC. Results were obtained from the Children's National Medical Center laboratory information system over the period January 5, 2001, to March 8, 2007. Patient results were accessed and used to establish reference intervals. All patient identifiers were removed except age and sex. Data was analyzed using the Hoffmann approach, and was computer adapted. Values are 2.5–97.5th percentiles.

INSULIN-LIKE GROWTH FACTOR-II (IGF-II)

Age	Male		Female	
	n	ng/mL µg/L	n	ng/mL µg/L
<3 y	10	310–6500	10	129–709
3–5 y	9	55–1091	15	124–928
6–8 y	15	237–977	20	145–1057
9–11 y	10	127–883	16	252–792
12–14 y	38	272–728	26	266–730
>15 y	14	317–745	14	176–756

Specimen Type(s)	Plasma
Reference(s)	Rosenfeld RG, Wilson DM, Lee PDK, et al. Insulin-like growth factors I and II in evaluation of growth retardation. J Ped 1986;109:428–33.
Method(s)	RIA after gel filtration on Sephadex G-50. See reference above.
Comment(s)	Subjects were normal healthy children, with heights between the 5th and 95th percentiles for age, selected from an outpatient pediatric clinic. Results are 2.5–97.5th percentiles.

IODIDE (URINE), (UI)

	Male			Female		
Age	n	UI Concentration µg/dL	UI/Creatinine Ratio (UI/CR) µg/g	n	UI Concentration µg/dL	UI/Creatinine Ratio (UI/CR) µg/g
6–9 y	1025	5.4–85.0	80–940	950	3.6–74.0	68–947
10–11 y	553	6.0–98.0	62–820	520	4.1–80.0	57–744
12–13 y	390	4.5–74.0	51–840	429	3.6–75.0	42–564
14–44 y	4782	2.4–66.0	34–420	5522	1.8–65.0	36–539

Female			
Age	n	UI Concentration µg/dL	UI/Creatinine Ratio (UI/CR) µg/dL
Child-bearing age (14–44 y)			
Pregnant	290	4.2–55.0	33–535
Not Pregnant	5232	1.8–65.0	36–541
All Females	5522	1.8–65.0	36–539

Specimen Type(s)	Urine
Reference(s)	Soldin OP, Soldin SJ, Pezzullo JC. Urinary iodine percentile ranges in the United States. Clin Chim Acta 2003;328:185–90.
Method(s)	Sandell-Koltoff reaction. Benotti J, Benotti N, Pino S, et al. Determination of the total iodine in urine, stool, diets, and tissue. Clin Chem 1965;11:932–6.
Comment(s)	The third National Health and Nutrition Examination Survey (NHANES III) (1988–1994) database of the civilian, non-institutionalized, iodide-sufficient US population was used. Gunter EW, Lewis BL, Konchikowski SM. Laboratory methods used for the Third National Health and Nutrition Examination Survey (NHANES III), 1988–1994. Hyattsville: Centers for Disease Control and Prevention, 1996.

		\multicolumn{6}{c}{**IRON**}					

		Male			Female		
Test	Age	n	µg/dL	µmol/L	n	µg/dL	µmol/L
1	1–5 y*	44*	22–136	4–25	44*	22–136	4–25
	6–9 y*	50*	39–136	7–25	50*	39–136	7–25
	10–14 y	31	28–134	5–24	40	45–145	8–26
	14–19 y	65	34–162	6–29	110	28–184	5–33
2	1–30 d	80	32–112	5.7–20.0	44	29–127	5.2–22.7
	31–365 d	87	27–109	4.8–19.5	75	25–126	4.5–22.6
	1–3 y	160	29–91	5.2–16.3	119	25–101	4.5–18.1
	4–6 y	114	25–115	4.5–20.6	88	28–93	5.0–16.7
	7–9 y	116	27–96	4.8–17.2	101	30–104	5.4–18.6
	10–12 y	107	28–112	5.0–20.0	91	32–104	5.7–18.6
	13–15 y	98	26–110	4.7–19.7	117	30–109	5.4–19.5
	16–18 y	92	27–138	4.8–24.7	99	33–102	5.9–18.3
3	0–90 d	44	72–203	12.9–36.3	32	75–235	13.4–42.1
	91 d–12 mo	54	23–142	4.1–25.4	38	60–192	10.7–34.4
	13–36 mo	66	25–126	4.5–22.6	56	55–162	9.8–29.0
	4–10 y	77	15–128	2.7–22.9	66	28–122	5.0–21.8
	11–14 y	59	32–107	5.7–19.2	57	25–102	4.5–18.3
	15–18 y	47	30–130	5.4–23.3	51	25–107	4.5–19.2

		5–11 am			5–11 pm		
Test	Age	n	µg/dL	µmol/L	n	µg/dL	µmol/L
4	0–24 mo	332	20–105	3.6–18.8	43	20–140	3.6–25.1
	2–9 y	370	20–105	3.6–18.8	100	20–145	3.6–26.0
	10–14 y	145	20–100	3.6–17.9	43	20–145	3.6–26.0
	15–18 y	125	20–100	3.6–17.9	42	20–145	3.6–26.0

		Male and Female		
Test	Age	n	µg/dL	µmol/L
5	0–12 mo	123	19.5–153.6	3.5–27.5
	1–5 y	147	9.5–150.8	1.7–27.0
	6–10 y	130	6.7–148.0	1.2–26.5
	11–14 y	150	19.0–156.3	3.4–28.0
	15–20 y	177	14.5–155.8	2.6–27.9

Specimen Type(s)	1	Plasma
	2–5	Plasma/serum
Reference(s)	1	Lockitch G, Halstead A, Wadsworth L, et al. Age- and sex-specific pediatric reference intervals for zinc, copper, selenium, iron, vitamins A and E, and related proteins. Clin Chem 1988;34:1625–8.
	2	Soldin SJ, Bailey J, Bjorn J, et al. Pediatric reference ranges for iron on the Hitachi 747 with Boehringer Mannheim reagents. Clin Chem 1999;45:A22. (Abstract)
	3	Soldin OP, Bierbower LH, Choi JJ, et al. Serum iron, ferritin, transferrin, total iron binding capacity, hs-CRP, LDL cholesterol and magnesium in children; new reference intervals using the Dade Dimension Clinical Chemistry System. Clin Chim Acta 2004;342:211–7.
	4	Ghoshal AK, Soldin SJ. Evaluation of the Dade Behring Dimension RxL: integrated chemistry system-pediatric reference ranges. Clin Chim Acta 2003;331:135–46.
	5	Chan MK, Seiden-Long I, Aytekin M, et al. Canadian Laboratory Initiative on Reference Interval Database (CALIPER): pediatric reference intervals for an integrated clinical chemistry and immunoassay analyzer, Abbott ARCHITECT ci8200. Clin Biochem 2009;42:885–91.
Method(s)	1	Ferrozine (Sigma Chemical Company). RA-100 Analyzer (Technicon Instruments, Tarrytown, NY).
	2	Hitachi 747 with Boehringer Mannheim reagents (Boehringer Mannheim Diagnostics, Indianapolis, IN).
	3	Siemens Dimension RxL Analyzer (Siemens Healthcare Diagnostics, Deerfield, IL).
	4	Under acidic conditions (pH 4.5), iron bound to the protein transferrin is released in the presence of the reducing agent, ascorbic acid. The resulting product, Fe^{++}, forms a blue complex with 3-(2-pyridyl)-5,6-bis-2-(5-furyl sulfonic acid)-1,2,4-triazine, disodium salt (Ferene®). Siemens Dimension RxL Analyzer (Siemens Healthcare Diagnostics, Deerfield, IL).
	5	Abbott Architect ci8200 (Abbott Diagnostics, Abbott Park, IL).
Comment(s)	1	The study population was healthy children. Non-parametric methods were used to determine the 0.025 and 0.975 fractiles. *No significant differences were found for males and females. These ranges were therefore derived from combined data.
	2–4	Study used hospitalized patients and a computerized approach adapted from the Hoffmann technique. Values are the 2.5–97.5th percentiles.
	5	1459 serum/plasma specimens from children attending select outpatient clinics were collected from the five age groups noted above. Values are 2.5–97.5th percentiles.

IRON-BINDING CAPACITY, TOTAL (TIBC)

Test	Age	Male			Female		
		n	µg/dL	µmol/L	n	µg/dL	µmol/L
1	1–5 y*	44	268–441	48–79	44	268–441	48–79
	6–9 y*	50	240–508	43–91	50	240–508	43–91
	10–14 y	31	302–508	54–91	40	318–575	57–103
	14–19 y	65	290–570	52–102	110	302–564	52–101
2	1–30 d	133	94–232	16.8–41.5	57	94–236	16.8–42.2
	31–182 d	78	116–322	20.8–57.6	69	89–311	15.9–55.7
	183–365 d	39	176–384	31.5–68.7	27	138–365	24.7–65.3
	1–3 y	131	204–382	36.5–68.3	103	184–377	32.9–67.5
	4–6 y	114	180–390	32.2–69.8	107	162–352	29.0–63.0
	7–9 y	116	183–369	32.8–66.1	113	167–336	29.9–60.1
	10–12 y	104	173–356	31.0–63.7	112	198–383	35.4–68.6
	13–15 y	106	193–377	34.5–67.5	143	169–358	30.3–64.1
	16–18 y	113	174–351	31.1–62.8	137	194–372	34.7–66.6
3	0–90 d	58	155–330	27.7–59.1	38	165–275	29.5–49.2
	91 d–12 mo	80	150–380	26.9–68.0	37	250–455	44.8–81.4
	13–36 mo	80	215–420	38.5–75.2	60	160–415	28.6–74.3
	4–10 y	79	185–415	33.1–74.3	72	260–385	46.5–68.9
	11–14 y	70	265–410	47.4–73.4	75	250–420	44.8–75.2
	15–18 y	40	270–415	48.3–74.3	41	285–410	51.0–73.4

Specimen Type(s)	1	Plasma
	2,3	Serum
Reference(s)	1	Lockitch G, Halstead A, Wadsworth L, et al. Age- and sex-specific pediatric reference intervals for zinc, copper, selenium, iron, vitamins A and E, and related proteins. Clin Chem 1988;34:1625–8.
	2	Soldin SJ, Hicks JM, Bailey J, et al. Pediatric reference ranges for total iron binding capacity and transferrin. Clin Chem 1997;43:S200. (Abstract)
	3	Soldin OP, Bierbower LH, Choi JJ, et al. Serum iron, ferritin, transferrin, total iron binding capacity, hs-CRP, LDL cholesterol and magnesium in children; new reference intervals using the Dade Dimension Clinical Chemistry System. Clin Chim Acta 2004;342:211–7.
Method(s)	1	Transferrin was measured by nephelometry. Behring LN with Behring kit (Behringwerke, Marburg, Germany). Calculated TIBC.
	2	Hitachi 747 using Boehringer Mannheim reagents (Boehringer Mannheim Diagnostics, Indianapolis, IN)
	3	Siemens Dimension RxL Analyzer (Siemens Healthcare Diagnostics, Deerfield, IL).
Comment(s)	1	Values are 2.5–97.5th percentiles. Transferrin iron-binding capacity calculated by multiplying the transferrin g/L $\times$ 23.1. *No significant differences were found for males and females. These ranges were therefore derived from combined data.
	2,3	Study used hospitalized patients and a computerized approach adapted from the Hoffmann technique. Values are 2.5–97.5th percentiles.

LACTATE

Male and Female

Test	Age	n	mg/dL	mmol/L
1	1–12 mo	12	10–21	1.1–2.3
	1–7 y	27	7–14	0.8–1.5
	7–15 y	9	5–8	0.6–0.9
2	0–90 d	85	30	3.3
	3–24 mo	150	28	3.1
	2–18 y	185	20	2.2
3	0–90 d	*	9–32	1.0–3.5
	3–24 mo		9–30	1.0–3.3
	2–18 y		9–22	1.0–2.4

Specimen Type(s)	1	Whole blood precipitated with perchloric acid.
	2	Plasma
	3	Plasma/serum
Reference(s)	1	Bonnefont JP, Specola NB, Vassault A, et al. The fasting test in pediatrics: application to pathological hypo- and hyperketotic states. Eur J Pediatr 1990;150:80–5.
	2	Soldin SJ, Baumel CR. Pediatric reference ranges for CSF protein and plasma lactate on the Vitros 500 analyzer. Clin Chem 2001:47:A108. (Abstract)
	3	Ghoshal AK, Soldin SJ. Evaluation of the Dade Behring Dimension RxL: integrated chemistry system-pediatric reference ranges. Clin Chim Acta 2003;331:135–46.
Method(s)	1	Standard enzymatic procedure. See reference.
	2	Vitros 500 analyzer using Ortho-Clinical Diagnostics reagents (Ortho-Clinical Diagnostics, Raritan, NJ).
	3	The lactic acid method employs the oxidation of lactate to pyruvate. Siemens Dimension RxL Analyzer (Siemens Healthcare Diagnostics, Deerfield, IL).
Comment(s)	1	The above results are 10–90th percentiles and refer to 15-h fasting values. For 20- and 24-h fasting values, see reference.
	2	Studies used hospitalized patients and a computerized approach adapted from the Hoffmann technique. Values are 97.5th percentile.
	3	*Numbers not provided. Reference ranges were obtained by comparing results from previously published data and using regression equations.

LACTATE DEHYDROGENASE (LDH)

Test	Age	Male n	Male U/L	Female n	Female U/L
1	1–30 d	119	550–2100	76	580–2000
	1–3 mo	108	480–1220	84	460–1150
	4–6 mo	100	400–1230	54	480–1150
	7–12 mo	97	380–1200	75	460–1060
2	1–3 y	50*	500–920	50*	500–920
	4–6 y	40*	470–900	40*	470–900
	7–9 y	80*	420–750	80*	420–750
	10–11 y	27	432–700	34	380–700
	12–13 y	31	470–750	49	380–640
	14–15 y	26	360–730	52	390–580
	16–19 y	40	340–670	61	340–670
3	1–30 d	77	125–735	65	145–765
	31–365 d	86	170–450	74	190–420
	1–3 y	135	155–345	109	165–395
	4–6 y	101	155–345	96	135–345
	7–9 y	125	145–300	104	140–280
	10–12 y	111	120–325	76	120–260
	13–15 y	105	120–290	101	100–275
	16–18 y	79	105–235	98	105–230
4	1–30 d	**	178–629	**	187–600
	1–3 mo		158–373		152–353
	4–6 mo		135–376		158–353
	7–12 mo		129–367		152–327
	1–3 y		164–286		164–286
	4–6 y		155–280		155–280
	7–9 y		141–237		141–237
	10–11 y		141–231		129–222
	12–13 y		141–231		129–205
	16–19 y		117–217		117–213

Specimen Type(s)	1,4	Plasma/serum
	2,3	Plasma
Reference(s)	1	Soldin SJ, Savwoir TV, Guo Y. Pediatric reference ranges for lactate dehydrogenase and uric acid during the first year of life on the Vitros 500 analyzer. Clin Chem 1997;43:S199. (Abstract)
	2	Lockitch G, Halstead AC, Albersheim S, et al. Age- and sex-specific pediatric reference intervals for biochemistry analytes as measured with the Ektachem-700 analyzer. Clin Chem 1988;34:1622–5.
	3	Soldin SJ, Bailey J, Bjorn S, et al. Pediatric reference ranges for LDH. Clin Chem 1995;41:S93. (Abstract)
	4	Ghoshal AK, Soldin SJ. Evaluation of the Dade Behring Dimension RxL: integrated chemistry system-pediatric reference ranges. Clin Chim Acta 2003;331:135–46.
Method(s)	1,2	Vitros 500 (1) and 700 (2) (Ortho-Clinical Diagnostics, Raritan, NJ).
	3	Measured on the Hitachi 747 using Boehringer Mannheim reagents (Boehringer Mannheim Diagnostics, Indianapolis, IN).
	4	The LDH method measures the oxidation of L-lactate to pyruvate with simultaneous reduction of nicotinamide adenine dinucleotide. Siemens Dimension RxL Analyzer (Siemens Healthcare Diagnostics, Deerfield, IL).
Comment(s)	1,3	Study used hospitalized patients and a computerized approach adapted from the Hoffmann technique to obtain the 2.5–97.5th percentiles.
	2	The study population was healthy children. Non-parametric methods were used to determine the 0.025 and 0.975 fractiles.
		*No significant differences were found for males and females. These ranges were therefore derived from combined data.
	4	**Numbers not provided.
		Reference ranges were obtained by comparing results from previously published data and using regression equations.

LACTATE/PYRUVATE RATIO

Male and Female

Age	n	Fed	15-h Fast
1 mo–1 y	12	—	10–28
1–7 y	27	12–18	11–18
7–15 y	9	8–20	8–20

Specimen Type(s)	See lactate and pyruvate.
Reference(s)	Hommes FA, ed. Techniques in diagnostic human biochemical genetics. New York, NY: Wiley-Liss, 1991:300.
Method(s)	See references for lactate and pyruvate.
Comment(s)	Results are 10–90th percentiles.

LEAD

	Male		Female	
Age	n	µg/dL	n	µg/dL
0–12 mo	1072	1.8–4.9	710	2.3–4.7
13–24 mo	593	1.7–5.6	455	2.2–5.2
2–4 y	2513	1.6–5.5	3004	2.2–5.1
5–9 y	1053	1.6–5.2	1025	2.5–4.8
10–11 y	232	1.3–4.9	161	1.9–4.5
12–18 y	121	1.8–4.4	207	2.1–4.4

Specimen Type(s)	EDTA whole blood
Reference(s)	Soldin OP, Hanak B, Soldin SJ. Blood lead concentrations in children: new ranges. Clin Chim Acta 2003;327:109–13.
Method(s)	Electrothermal atomization atomic absorption spectrophotometric assay, AAnalyst 600 (Perkin Elmer, Wilton, CT).
Comment(s)	CDC guidelines place the threshold for concern at 10 µg/dL.

LEPTIN

	Male		Female	
Age	n	µg/L	n	µg/L
Maternal serum	360	59.1	353	66.0
Cord blood	361	33.6	348	42.7

Specimen Type(s)	Serum
Reference(s)	Weyermann M, Beermann C, Brenner H, Rothenbacher D. Adiponectin and leptin in maternal serum, cord blood, and breast milk. Clin Chem 2006;52:2095–102.
Method(s)	Leptin was measured with a commercially available ELISA (Rand D Systems, Minneapolis, MN).
Comment(s)	Values are 97.5th percentiles.

LIPASE

Test	Age	Male and Female	
		n	U/L
1	1–30 d	83	6–55
	31–182 d	121	4–29
	183–365 d	56	4–23
	1–3 y	148	4–31
	4–9 y	96	3–32
	10–18 y	142	4–29
2	0–12 mo	94	4–41
	1–5 y	148	5–63
	6–10 y	126	5–51
	11–14 y	160	4–43
	15–20 y	145	4–43

Test	Age	Male		Female	
		n	U/L	n	U/L
3	0–90 d	39*	10–85	39*	10–85
	3–12 mo	113	13–95	137	9–128
	1–<2 y	118	15–135	222	15–150
	2–<7 y	323	15–175	243	10–150
	7–<11 y	216	10–175	209	13–150
	11–<15 y	75	10–195	150	10–180
	15–18 y	70	10–195	79	10–220
4	0–90 d	**	145–174	**	145–174
	3–12 mo		146–178		144–190
	1–<2 y		147–193		147–199
	2–<7 y		147–209		145–199
	7–<11 y		145–209		146–199
	11–<15 y		145–216		145–211
	15–18 y		145–216		145–226

Specimen Type(s)	1–4	Plasma/serum
Reference(s)	1	Soldin SJ, Bailey J., Beatey J., et al. Pediatric reference ranges for lipase. Clin Chem 1995;41:S93. (Abstract)
	2	Chan MK, Seiden-Long I, Aytekin M, et al. Canadian Laboratory Initiative on Reference Interval Database (CALIPER): pediatric reference intervals for an integrated clinical chemistry and immunoassay analyzer, Abbott ARCHITECT ci8200. Clin Biochem 2009;42:885–91. Values rounded to the nearest whole number.
	3	Soldin SJ, Ojeifo O. Pediatric reference ranges for lipase. Clin Chem 1999;45:A22. (Abstract)
	4	Ghoshal AK, Soldin SJ. Evaluation of the Dade Behring Dimension RxL: integrated chemistry system-pediatric reference ranges. Clin Chim Acta 2003;331:135–46.
Method(s)	1	Hitachi 717 using Sigma reagents (Boehringer Mannheim Diagnostics, Indianapolis, IN).
	2	Abbott Architect ci8200 (Abbott Diagnostics, Abbott Park, IL).
	3	Vitros 500 using Vitros reagents (Ortho-Clinical Diagnostics, Raritan, NJ).
	4	The LIP method technology uses a glycerol derivative esterified in the 1-position with a dicarbonic acid-resorufin ester as the substrate. Lipase activity is maximized in the presence of colipase, calcium chloride, and bile salt. Siemens Dimension RxL Analyzer (Siemens Healthcare Diagnostics, Deerfield, IL).
Comment(s)	1	Study used hospitalized patients and a computerized approach to removing outliers. Values are 2.5–97.5th percentiles.
	2	1459 serum/plasma specimens from children attending select outpatient clinics were collected from the five age groups noted above. Values are 2.5–97.5th percentiles..
	3	Study used hospitalized patients and a computerized approach to removing outliers. Values are 2.5–97.5th percentiles. *Male and female date combined.
	4	**Numbers not provided. Reference ranges were obtained by comparing results from previously published data and using regression equations.

LOW-DENSITY LIPOPROTEIN-CHOLESTEROL (LDL-C)

Test	Age	n	mmol/L	mg/dL
			Male and Female	
1	2–12 mo	100	0.8–3.0	32–117
	2–10 y	120	1.0–3.6	38–140
2	0–12 mo	129	0.5–3.9	17–151
	1–5 y	152	1.5–4.0	59–153
	6–10 y	133	1.5–3.7	58–143
	11–14 y	154	1.4–4.2	56–163
	15–20 y	182	1.4–4.2	52–164
3	0–<1 y	99	0.8–3.5	31–135
	≥1–≤20 y	481	1.3–4.0	50–154

Test	Age		Male			Female	
		n	mmol/L	mg/dL	n	mmol/L	mg/dL
4	5–9 y	*	1.6–3.3	63–129	*	1.8–3.6	68–140
	10–14 y		1.7–3.4	64–133		1.8–3.5	68–136
	15–19 y		1.6–3.4	62–130		1.5–3.6	59–137
5	0–90 d	61	0.5–2.2	20–83	45	0.4–2.5	15–95
	91 d–12 mo	69	0.9–3.1	35–120	39	1.2–3.2	45–125
	13–36 mo	82	0.9–3.2	35–125	67	0.9–3.2	35–125
	4–10 y	35	1.2–3.6	45–140	83	0.9–3.5	35–135
	11–15 y	75	1.2–3.1	45–120	69	1.3–3.4	50–130
	16–18 y	55	1.4–3.1	55–120	48	1.8–3.1	70–120

Specimen Type(s)	1,4	Serum
	2,3,5	Plasma/serum
Reference(s)	1	Baroni S, Scribano D, Valentini P, et al. Serum apolipoprotein A1, B, CII, CIII, E, and lipoprotein (a) levels in children. Clin Biochem 1996;29:603–5.
	2	Chan MK, Seiden-Long I, Aytekin M, et al. Canadian Laboratory Initiative on Reference Interval Database (CALIPER): pediatric reference intervals for an integrated clinical chemistry and immunoassay analyzer, Abbott ARCHITECT ci8200. Clin Biochem 2009;42:885–91.
	3	Kulasingam V, Jung BP, Blasutig IM, et al. Pediatric reference intervals for 28 chemistries and immunoassays on the Roche cobas® 6000 analyzer—A CALIPER pilot study. Clin Biochem 2010;43;1045–50.
	4	Soldin SJ, Rifai N, Hicks JM, eds. Biochemical basis of pediatric disease, 3rd ed. Washington, DC: AACC Press, 1998:468.
	5	Soldin OP, Bierbower LH, Choi JJ, et al. Serum iron, ferritin, transferrin, total iron binding capacity, hs-CRP, LDL cholesterol and magnesium in children; new reference intervals using the Dade Dimension Clinical Chemistry System. Clin Chim Acta 2004;342:211–7.
Method(s)	1	Calculated as described in Friedewald WT, Levy RI, Frederickson DC. Estimation of the concentration of low-density lipoprotein cholesterol in plasma, without use of the preparative ultracentrifuge. Clin Chem 1972;18:499–502.
	2	Abbott Architect ci8200 (Abbott Diagnostics, Abbott Park, IL).
	3	Roche cobas® 6000 analyzer (Roche Diagnostics Limited, West Sussex, UK).
	4	See references.
	5	Siemens Dimension RxL Analyzer (Siemens Healthcare Diagnostics, Deerfield, IL).
Comment(s)	1	Results are the 2.5–97.5th percentiles for healthy children.
	2	1459 serum/plasma specimens from children attending select outpatient clinics were collected from the five age groups noted above. Values are 2.5–97.5th percentiles.
	3	Approximately 600 outpatient samples from a pediatric population deemed to be metabolically stable were subdivided into five age classes ranging from 0 to 20 years of age and further partitioned by gender. Values are 2.5–97.5th percentiles.
	4	*Numbers of participants not provided. Results are 5–95th percentiles.
	5	Study used hospitalized patients and a computerized approach adapted from the Hoffmann technique. Values are the 2.5–97.5th percentiles.

LUTEINIZING HORMONE (LH)

Test	Age	Male n	Male U/L	Female n	Female U/L
1	<2 y	179	0.5–1.9	33	<0.5
	2–5 y	138	<0.5	94	<0.5
	6–10 y	96	<0.5	136	<0.5
	11–20 y	89	0.5–5.3	263	0.5–9.0
2	<2 y	179	1.3–2.8	33	<1.3
	2–5 y	138	<1.3	94	<1.3
	6–10 y	96	<1.3	136	<1.3
	11–20 y	89	1.3–5.7	263	1.3–10.2

Test	Male Age	Male n	Male U/L	Female Age	Female n	Female U/L
3	0–<13 y	47	<0.1–4.0	0–<6 y	51	<0.1–3.3
	13–<19 y	62	<0.1–3.7	6–<11 y	106	<0.1–5.0
				11–<15 y	87	<0.1–13.4
				15–<19 y	111	<0.1–16.4

Test	Age	n	Male mIU/mL	Female mIU/mL
4	0–7 d	**	1.5–5.0	1.5–5.0
	2 wk–1 y		Increases by 2 weeks to 3.0–22.0. Declines to prepubertal levels by 1 year.	Increases by 2 weeks to 1.8–13.0. Declines to prepubertal levels by 1 year.
	Prepubertal children		1.0–3.5	1.0–3.5
	Pubertal children		Rise to adult values: 3.0–10.0	Rise to adult values: Follicular 3.0–11.0 Mid-cycle 18.0–70.0 Luteal 1.0–11.0

Specimen Type(s)	1–3	Plasma/serum
	4	Serum
Reference(s)	1	Soldin SJ, Morales A, Albalos F, et al. Pediatric reference ranges on the Abbott IMx for FSH, LH, prolactin, TSH, T_4, T_3, free T_4, free T_4, T-uptake, IgE and ferritin. Clin Biochem 1995;28:603–6.
	2	Murthy JN, Hicks JM, Soldin SJ. Evaluation of the Technicon Immuno I Random Access Immunoassay Analyzer and calculation of pediatric reference ranges for endocrine tests, T-uptake and ferritin. Clin Biochem 1995;28:181–5.
	3	Soldin OP, Hoffman EG, Waring MA, Soldin SJ. Pediatric reference intervals for FSH, LH, estradiol, T3, free T3, cortisol, and growth hormone on the DPC IMMULITE 1000. Clin Chim Acta 2005;355:205–10.
	4	Dugaw KA, Jack RM, Rutledge J. Pediatric reference ranges for FSH and LH on the Vitros ECi analyzer. Clin Chem 2001;47:A108. (Abstract)
Method(s)	1	IMx (Abbott Laboratories, Abbott Park, IL).
	2	Immuno I (Bayer Corp., Tarrytown, NY).
	3	Siemens IMMULITE® 1000 analyzer (Siemens Healthcare Diagnostics, Deerfield, IL).
	4	Chemiluminescent immunoassay, Vitros ECi (Ortho-Clinical Diagnostics, Raritan, NJ).
Comment(s)	1,2	Study used hospitalized patients and a computerized approach to removing outliers. Values are 2.5–97.5th percentiles.
	3	Study used hospitalized patients and a computerized approach adapted from the Hoffmann technique. Values are the 2.5–97.5th percentiles.
	4	**Ages ranged from 1 h to 18 y with a total of 92 specimens.

		Male			Female		
		\multicolumn{6}{c}{**MAGNESIUM**}		
Test	Age	n	mg/dL	mmol/L	n	mg/dL	mmol/L
1	1–30 d	68	1.7–2.4	0.70–0.99	18	1.7–2.5	0.70–1.03
	31–365 d	62	1.6–2.5	0.66–1.03	37	1.9–2.4	0.78–0.99
	1–3 y	140	1.7–2.4	0.70–0.99	103	1.7–2.4	0.70–0.99
	4–6 y	120	1.7–2.4	0.70–0.99	63	1.7–2.2	0.70–0.91
	7–9 y	118	1.7–2.3	0.70–0.95	102	1.6–2.3	0.66–0.95
	10–12 y	111	1.6–2.2	0.66–0.91	74	1.6–2.2	0.66–0.91
	13–15 y	84	1.6–2.3	0.66–0.95	113	1.6–2.3	0.66–0.95
	16–18 y	73	1.5–2.2	0.62–0.91	92	1.5–2.2	0.62–0.91
2	Premature 0–6 d	48	1.6–2.7	0.65–1.10	48	1.6–2.7	0.65–1.10
	Newborn 0–6 d	134	1.2–2.6	0.48–1.05	134	1.2–2.6	0.48–1.05
	Premature 7–30 d	24	1.8–2.4	0.75–1.00	24	1.8–2.4	0.75–1.00
	Newborn 7–30 d	89	1.6–2.4	0.65–1.00	89	1.6–2.4	0.65–1.00
	1 mo–1 y	164	1.6–2.6	0.65–1.05	164	1.6–2.6	0.65–1.05
	1–2 y	102	1.6–2.6	0.65–1.05	102	1.6–2.6	0.65–1.05
	2–6 y	138	1.5–2.4	0.60–1.00	138	1.5–2.4	0.60–1.00
	6–10 y	130	1.6–2.3	0.65–0.95	130	1.6–2.3	0.65–0.95
	10–14 y	133	1.6–2.2	0.65–0.90	133	1.6–2.2	0.65–0.90
	14 y +	139	1.5–2.3	0.60–0.95	139	1.5–2.3	0.60–0.95
3	0–90 d	59	1.45–2.15	0.59–0.88	45	1.49–2.05	0.61–0.84
	91 d–12 mo	69	1.59–2.49	0.65–1.02	38	1.60–2.20	0.66–0.90
	13–36 mo	82	1.59–2.20	0.65–0.90	65	1.51–2.20	0.62–0.90
	4–10 y	94	1.49–2.20	0.61–0.90	84	1.60–2.50	0.66–1.03
	11–15 y	89	1.35–2.05	0.55–0.84	82	1.60–2.09	0.66–0.86
	16–18 y	42	1.55–2.10	0.64–0.86	48	1.49–1.90	0.61–0.78

		Male and Female		
Test	Age	n	mg/dL	mmol/L
4	0–12 mo	128	1.9–3.2	0.8–1.3
	1–5 y	155	1.9–3.4	0.8–1.4
	6–10 y	134	1.9–3.6	0.8–1.5
	11–14 y	154	1.9–2.4	0.8–1.0
	15–20 y	183	1.9–2.4	0.8–1.0

Specimen Type(s)	1,2	Plasma
	3,4	Plasma/serum
Reference(s)	1	Hicks JM, Bailey J, Bjorn S, et al. Pediatric reference ranges for plasma magnesium. Clin Chem 1995;41:S93. (Abstract)
	2	Meites S, ed. Pediatric clinical chemistry, 3rd ed. Washington, DC: AACC Press, 1989:191.
	3	Soldin OP, Bierbower LH, Choi JJ, et al. Serum iron, ferritin, transferrin, total iron binding capacity, hs-CRP, LDL cholesterol and magnesium in children; new reference intervals using the Dade Dimension Clinical Chemistry System. Clin Chim Acta 2004;342:211–7.
	4	Chan MK, Seiden-Long I, Aytekin M, et al. Canadian Laboratory Initiative on Reference Interval Database (CALIPER): pediatric reference intervals for an integrated clinical chemistry and immunoassay analyzer, Abbott ARCHITECT ci8200. Clin Biochem 2009;42:885–91.
Method(s)	1	Magnesium reacts with calmagite to form a reddish violet chromophore. Hitachi 747 with Boehringer Mannheim reagents (Boehringer Mannheim Diagnostics, Indianapolis, IN).
	2	Colorimetric using a formazan dye. Vitros 700 (Ortho-Clinical Diagnostics, Raritan, NJ).
	3	Methylthymol blue forms a blue complex with magnesium. Calcium interference is minimized by forming a complex between calcium and Ba-EGTA (chelating agent). Siemens Dimension RxL Analyzer (Siemens Healthcare Diagnostics, Deerfield, IL).
	4	Abbott Architect ci8200 (Abbott Diagnostics, Abbott Park, IL).
Comment(s)	1	Study used hospitalized patients and a computerized approach adapted from the Hoffmann technique to obtain 2.5–97.5th percentiles.
	2	Results are 5–95th percentiles from patients at Children's Hospital, Columbus, OH. Males and females studied together.
	3	Study used hospitalized patients and a computerized approach adapted from the Hoffmann technique to obtain the 2.5–97.5th percentiles.
	4	1459 serum/plasma specimens from children attending select outpatient clinics were collected from the five age groups noted above. Values are 2.5–97.5th percentiles.

MANGANESE

Male and Female

Age	n	µg/L	nmol/L
0–<0.5 y	13	0.5–3.8	8.7–68.7
0.5–<1 y	18	0.6–3.7	10.7–66.3
1–<2 y	15	0.0–3.4	0.0–61.1
2–<4 y	23	0.4–2.3	8.1–42.1
4–<6 y	19	0.2–2.7	3.3–49.3
6–<10 y	25	0.3–2.3	5.2–40.8
10–<14 y	21	0.4–2.0	7.6–36.4
14–<18 y	17	0.4–1.5	7.8–27.8

Specimen Type(s)	Plasma/serum
Reference(s)	Rükgauer M, Klein J, Kruse-James JD. Reference values for the trace elements copper, manganese, selenium, and zinc in the serum/plasma of children, adolescents, and adults. J Trace Elements Med Biol 1997;11:92–8.
Method(s)	Atomic absorption spectrophotometry with Zeeman background compensation. Perkin Elased ETAAS, Zeeman 3030 (Uberlingen, Germany).
Comment(s)	Study population was drawn from patients visiting the outpatient department or surgical or orthopedic ward for preoperative workup. Results are mean ±2 SDs. (2.5th–97.5th percentiles)

METANEPHRINE (URINE)

Male and Female

Age	n	µg/24h	µmol/24h	µg/g creatinine
3–8 y	76	5–113	0.025–0.514	47–240
9–12 y	44	21–154	0.107–0.782	40–220
13–17 y	43	32–167	0.162–0.848	33–145
Adults	51	45–290	0.228–1.472	31–140

Specimen Type(s)	Urine
Reference(s)	Pediatric endocrine testing. Nichols Institute, 1993:30.
Method(s)	HPLC electrochemical detection.
Comment(s)	Results are 2.5–97.5th percentiles.

METHYLMALONIC ACID

Male and Female

Age	n	µmol/L
4–8 y	19	0.06–0.24
8–14 y	20	0.03–0.26

Specimen Type(s)	Serum
Reference(s)	Straczek J, Felden F, Dousset B. Quantification of methylmalonic acid in serum measured by capillary gas chromatography-mass spectrometry as tertbutyldimethylsilyl derivates. J Chromat 1993;620:1–7.
Method(s)	Gas chromatography-mass spectrometry.
Comment(s)	Results are 2.5–97.5th percentiles.

β_2-MICROGLOBULIN

	Male		Female	
Age	n	µg/L	n	µg/L
1–30 d	68	1603–4790	50	1722–4547
31–182 d	73	1423–3324	68	1024–3774
183–365 d	39	897–3095	27	999–2282
1–3 y	158	827–2228	129	742–2396
4–6 y	142	567–2260	123	546–2170
7–9 y	97	772–1712	79	736–1766
10–12 y	92	699–1836	68	704–1951
13–15 y	103	681–1954	82	787–1916
16–18 y	54	724–1874	77	555–1852

Specimen Type(s)	Plasma/serum
Reference(s)	Soldin SJ, Hicks JM, Bailey J, et al. Pediatric reference ranges for β2-microglobulin and ceruloplasmin. Clin Chem 1997;43:S199. (Abstract)
Method(s)	RIA. Pharmacia β2-Micro RIA (Pharmacia, Uppsala, Sweden).
Comment(s)	Study used hospitalized patients and a computerized approach adapted from the Hoffmann technique. Values are the 2.5–97.5th percentiles.

MUCOPOLYSACCHARIDES (URINE)

Male and Female		
Age	n	mg/mmol creatinine
0–<6 mo	102	15.2–52.0
6 mo–<12 mo	34	15.1–31.5
1–<2 y	43	9.1–29.9
2–<4 y	55	7.7–21.3
4–<6 y	45	7.6–14.4
6–<8 y	30	5.7–12.9
8–<10 y	16	5.2–11.6
10–<15 y	27	2.4–10.6
15–<20 y	10	1.5–6.7
>20 y	31	1.5–5.1

Specimen Type(s)	Urine
Reference(s)	de Jong JGN, Wevers RA, Liebrand-van Sambeek R. Measuring urinary glycosaminoglycans in the presence of protein: An improved screening procedure for mucopolysaccharides based on dimethyl-methylene blue. Clin Chem 1992;36:803–7.
Method(s)	Dimethylene blue-Tris assay.
Comment(s)	Results are the 2.5–97.5th percentiles

NOREPINEPHRINE/NORADRENALINE (PLASMA)

Male and Female

Test	Age	n	nmol/L	pg/mL
1	2–10 d	21	1.0–7.0	169–1184
	10 d–3 mo	10	2.2–12.3	372–2081
	3 mo–1 y	14	1.6–6.6	271–1117
	1–2 y	13	0.4–10.7	68–1810
	2–3 y	8	1.1–8.7	186–1472
	3–15 y	20	0.5–7.4	85–1252
2	30 min after birth	16	2.5–9.1	422–1538
	2 h after birth	16	3.1–6.5	532–1108
	3 h after birth	16	0.7–8.5	112–1436
	12 h after birth	16	1.2–1.9	195–327
	24 h after birth	16	1.5–2.8	250–470
	48 h after birth	16	1.3–2.0	221–345

Specimen Type(s)	1,2	Plasma
Reference(s)	1	Candito M, Albertini M, Politano S, et al. Plasma catecholamine levels in children. J Chrom Biomed Appl 1993;617:304–7.
	2	Eliot RJ, Lam R, Leake RD, et al. Plasma catecholamine concentrations in infants at birth and during the first 48 hours of life. J Pediatrics 1980;96:311–5.
Method(s)	1	High performance liquid chromatography.
	2	Radioenzymatic method.
Comment(s)	1	Study population consisted of 86 healthy children (62 males, 24 females) aged 2 days to 15 years.
	2	Study performed on 16 term vaginally delivered infants. Results are mean ±2 SDs.

NOREPINEPHRINE/NORADRENALINE (URINE)

		Male and Female		
Test	Age	n	µg/g creatinine	mmol/mol creatinine
1	0–24 mo	24	<420	<0.280
	2–4 y	37	<120	<0.080
	5–9 y	40	<89	<0.059
	10–19 y	41	<82	<0.055
2	<1 y	18	25–310	0.017–0.207
	1–4 y	24	25–290	0.017–0.194
	4–10 y	23	27–108	0.018–0.072
	10–18 y	20	4–105	0.003–0.070

Specimen Type(s)	1,2	Urine
Reference(s)	1	Soldin SJ, Lam G, Pollard A, et al. High performance liquid chromatographic analysis of urinary catecholamines employing amperometric detection: Reference values and use in laboratory diagnosis of neural crest tumors. Clin Biochem 1980;13:285–91.
	2	Rosano TG. Liquid-chromatographic evaluation of age-related changes in the urinary excretion of free catecholamines in pediatric patients. Clin Chem 1984;30:301–3.
Method(s)	1,2	HPLC with electrochemical detection.
Comment(s)	1	Study involved healthy children. Results quoted are 95th percentile.
	2	Urine was obtained from 85 pediatric patients (diagnosis of neoplasia excluded). Results are 0–100th percentiles.

NORMETANEPHRINE (URINE)

	Male and Female			
Age	n	µg/24h	µmol/24h	µg/g creatinine
3–8 y	76	13–252	0.071–1.375	62–705
9–12 y	44	32–346	0.175–1.888	81–583
13–17 y	43	63–402	0.344–2.194	95–375
Adults	51	82–500	0.448–2.729	47–310

Specimen Type(s)	Urine
Reference(s)	Pediatric endocrine testing. Nichols Institute, 1993:30.
Method(s)	HPLC electrochemical detection.
Comment(s)	Results are 2.5–97.5th percentiles.

OSMOLALITY

Test	Age	n	mOsm/kg
	Male and Female		
1	Birth	44	275–300
	7 d	17	276–305
	28 d	17	274–305
2	1 mo–18 y	*	280–300

Specimen Type(s)	1,2	Serum
Reference(s)	1	Davies DP. Plasma osmolality and protein intake in preterm infants. Arch Dis Child 1973;48:575–9.
		Meites S, ed. Pediatric clinical chemistry, 3rd ed. Washington, DC: AACC Press, 1989:202–3.
	2	Children's National Medical Center. Unpublished data.
Method(s)	1,2	Freezing-point depression.
Comment(s)	1	The report is based on Davies' study of 53 preterm infants with birthweights <2.5 kg.
	2	*Numbers not provided.

OXYGEN, PARTIAL PRESSURE (PO$_2$)

	Male and Female			
Test	Age	n	mmHg	kPa
1	Birth	*	8–24	1.1–3.2
	5–10 min		33–75	4.4–10.0
	30 min		31–85	4.1–11.3
	>1 h		55–80	7.3–10.6
	1 d		54–95	7.2–12.6
	>1 d		83–108	11.0–14.4
2	Premature neonates	248	31–57	4.1–7.6

Specimen Type(s)	1	Arterial whole blood
	2	Capillary blood
Reference(s)	1	Behrman RE, ed. Nelson textbook of pediatrics, 14th ed. Philadelphia, PA: WB Saunders Company, 1992:1818.
	2	Soldin SJ. Children's National Medical Center. Unpublished data.
Method(s)	1	Not given.
	2	I-Stat (Abbott Laboratories, Abbott Park, IL).
Comment(s)	1	*Numbers not provided.
	2	Study used hospitalized patients and a computerized approach adapted from the Hoffmann technique.

OXYGEN SATURATION

		Male and Female		
Test	Age	n	%	Saturated Fraction
1	Newborn	*	85–90	0.85–0.90
	Thereafter		95–99	0.95–0.99
2	Premature neonates	248	52–90	0.52–0.90

Specimen Type(s)	1	Arterial whole blood
	2	Capillary blood
Reference(s)	1	Behrman RE, ed. Nelson textbook of pediatrics, 14th ed. Philadelphia, PA: WB Saunders Company, 1992:1818.
	2	Soldin SJ. Children's National Medical Center. Unpublished data.
Method(s)	1	Not given.
	2	I-Stat (Abbott Laboratories, Abbott Park, IL).
Comment(s)	1	*Numbers not provided.
	2	Study used hospitalized patients and a computerized approach adapted from the Hoffmann technique.

PARATHYROID HORMONE (PTH)

Test	Male and Female		
	Age	n	pg/mL
1	Mid-Molecule (C-Terminal)		
	1–16 y	39	51–217
	N-Terminal Specific		
	2–13 y	18	14–21
	Intact		
	Cord blood	20	≤3.0
	2–20 y	150	9–52
2	Intact		
	0–<1 y	67	10–112
	1–<3 y	62	10–71
	3–<10 y	150	12–120
	10–<13 y	82	13–85
	13–<15 y	76	10–117
	15–18 y	136	13–100

Specimen Type(s)	1	Serum
	2	Plasma/serum
Reference(s)	1	Pediatric endocrine testing. Nichols Institute, 1993:31–2.
	2	Soldin OP, Dahlin JR, Gresham EG, et al. IMMULITE 2000 age and sex-specific reference intervals for alpha fetoprotein, homocysteine, insulin, insulin-like growth factor-1, insulin-like growth factor binding protein-3, C-peptide, immunoglobulin E and intact parathyroid hormone. Clin Biochem 2008;41:937–42.
Method(s)	1	Intact (immunoradiometric assay), N-terminal specific (radioimmunoassay), C-terminal (radioimmunoassay).
	2	IMMULITE® 2000 immunoassay system. (Siemens Healthcare Diagnostics, Deerfield, IL).
Comment(s)	1	Results are 2.5–97.5th percentiles.
	2	Study was conducted at both Children's National Medical Center and Georgetown University, Washington, DC. Results were obtained from the Children's National Medical Center laboratory information system over the period January 5, 2001, to March 8, 2007. Patient results were accessed and used to establish reference intervals. All patient identifiers were removed except age and sex. Data was analyzed using the Hoffmann approach, and was computer adapted. Values are 2.5–97.5th percentiles.

PEPTIDE YY (PYY)		
Male and Female		
Age	n	ng/L
28–36 wk gestation	62	695–1157
37–41 wk gestation	15	356–1294

Specimen Type(s)	Serum
Reference(s)	Siahanidou T, Mandyla H, Vounatsou M, Anagnostakis D, Papassotiriou I, Chrousos GP. Circulating peptide YY concentrations are higher in preterm than full-term infants and correlate negatively with body weight and positively with serum ghrelin concentrations. Clin Chem 2005;51:2131–7.
Method(s)	Serum PYY was measured with a human PYY RIA (Linco Research, St. Charles, MO).
Comment(s)	Values are 2.5–97.5th percentiles.

	pH		
Test	**Age**	**n**	**Male and Female**
1	0–1 mo	200	7.18–7.51
	1–6 mo	100	7.18–7.50
	6–12 mo	85	7.27–7.49
2	Cord blood[a]	169	7.26–7.50
	2–5 d[b]	28	7.30–7.49
3	Premature neonates	248	7.23–7.43
Specimen Type(s)	1,2	Whole blood	
	3	Capillary blood	
Reference(s)	1	Snell J, Greeley C, Colaco A, et al. Pediatric reference ranges for arterial pH, whole blood electrolytes and glucose. Clin Chem 1993;39:1173. (Abstract)	
	2	[a]Dickman KA. In: Meites S, ed. Pediatric clinical chemistry, 3rd ed. Washington, DC: AACC Press, 1989:24–5.	
		[b]Lockitch G, Halstead AC. In: Meites S, ed. Pediatric clinical chemistry, 3rd ed. Washington, DC: AACC Press, 1989:25.	
	3	Soldin SJ. Children's National Medical Center. Unpublished data.	
Method(s)	1	Ion specific electrode. (Ciba Corning 288 blood gas system. Ciba Corning Diagnostics, East Walpole, MA).	
	2	[a]Ion specific electrode. (IL-813, Instrumentation Laboratories, Lexington, MA).	
		[b]Ion specific electrode. (Nova Biomedical Stat Profile, Nova Biomedical, Waltham, MA).	
	3	I-Stat (Abbott Laboratories, Abbott Park, IL).	
Comment(s)	1	Study used hospitalized patients and a computerized approach to removing outliers. Values are 2.5–97.5th percentiles.	
	2	[a]In-house study of newborns with Apgar scores of 7 or more.	
		[b]Healthy term infants. Mean birth weight 3465 g. Results are 2.5–97.5th percentiles.	
	3	Study used hospitalized patients and a computerized approach adapted from the Hoffmann technique.	

PHOSPHORUS

Test	Age	Male n	Male mg/dL	Male mmol/L	Female n	Female mg/dL	Female mmol/L
1	1–30 d	62	3.9–6.9	1.25–2.25	66	4.3–7.7	1.40–2.50
	31–365 d	83	3.5–6.6	1.15–2.15	66	3.7–6.5	1.20–2.10
	1–3 y	126	3.1–6.0	1.00–1.95	119	3.4–6.0	1.10–1.95
	4–6 y	112	3.3–5.6	1.05–1.80	107	3.2–5.5	1.05–1.80
	7–9 y	117	3.0–5.4	0.95–1.75	107	3.1–5.5	1.00–1.80
	10–12 y	135	3.2–5.7	1.05–1.85	115	3.3–5.3	1.05–1.70
	13–15 y	109	2.9–5.1	0.95–1.65	110	2.8–4.8	0.90–1.55
	16–18 y	95	2.7–4.9	0.85–1.60	122	2.5–4.8	0.80–1.55
2	0–5 d (<2.5 kg)	50	4.6–8.0	1.50–2.60	50	4.6–8.0	1.50–2.60
	1–3 y	50	3.9–6.5	1.25–2.10	50	3.9–6.5	1.25–2.10
	4–6 y	38	4.0–5.4	1.30–1.75	38	4.0–5.4	1.30–1.75
	7–9 y	72	3.7–5.6	1.20–1.80	72	3.7–5.6	1.20–1.80
	10–11 y	62	3.7–5.6	1.20–1.80	62	3.7–5.6	1.20–1.80
	12–13 y	73	3.3–5.4	1.05–1.75	73	3.3–5.4	1.05–1.75
	14–15 y	91	2.9–5.4	0.95–1.75	91	2.9–5.4	0.95–1.75
	16–19 y	107	2.8–4.6	0.90–1.50	107	2.8–4.6	0.90–1.50
3	0–30 d	181	2.7–7.2	0.87–2.33	140	3.0–8.0	0.97–2.58
	31–90 d	84	3.0–6.8	0.97–2.20	87	3.0–7.5	0.97–2.42
	3–12 mo	109	3.0–6.9	0.97–2.23	119	2.5–7.0	0.81–2.26
	13–24 mo	69	2.5–6.4	0.81–2.07	78	3.0–6.5	0.97–2.10
	2–<13 y	148	3.0–6.0	0.97–1.94	254	2.5–6.0	0.81–1.94
	13–<16 y	175	3.0–5.4	0.97–1.74	72	3.0–5.6	0.97–1.81
	16–<18 y	72	3.0–5.2	0.97–1.68	196	3.0–4.8	0.97–1.55
4	0–30 d	*	2.8–7.0	0.90–2.26	*	3.1–7.7	1.00–2.49
	31–90 d		3.1–6.6	1.00–2.13		3.1–7.2	1.00–2.32
	3–12 mo		3.1–6.6	1.00–2.13		3.1–6.8	1.00–2.20
	13–24 mo		3.1–6.2	1.00–2.00		3.1–6.3	1.00–2.03
	2–<13 y		3.1–5.9	1.00–1.90		3.1–5.9	1.00–1.90
	13–<16 y		3.1–5.3	1.00–1.71		3.1–5.5	1.00–1.78
	16–<18 y		3.1–5.1	1.00–1.65		3.1–4.8	1.00–1.55

Test	Male and Female				
	Age	n	mg/dL	mmol/L	
5	0–12 mo	129	4.8–7.8	1.54–2.52	
	1–5 y	155	4.4–6.8	1.43–2.19	
	6–10 y	134	3.8–6.2	1.22–2.00	
		Male			Female		
	Age	n	mg/dL	mmol/L	n	mg/dL	mmol/L
	11–14 y	42	3.8–6.1	1.21–1.96	111	3.0–5.8	0.97–1.86
	15–20 y	35	2.7–5.4	0.86–1.73	147	3.3–5.0	1.06–1.61

Specimen Type(s)	1–5	Plasma/serum
Reference(s)	1	Soldin SJ, Hicks JM, Bailey J, et al. Pediatric reference ranges for phosphate on the Hitachi 747 analyzer. Clin Chem 1997;43:S198. (Abstract)
	2	Lockitch G, Halstead AC, Albersheim S, et al. Age- and sex-specific pediatric reference intervals for biochemistry analytes as measured with the Ektachem-700 analyzer. Clin Chem 1988;34:1622–5.
	3	Soldin SJ, Hunt C, Hicks JM. Pediatric reference ranges for phosphorus on the Vitros 500 Analyzer. Clin Chem 1999;45:A22. (Abstract)
	4	Ghoshal AK, Soldin SJ. Evaluation of the Dade Behring Dimension RxL: integrated chemistry system-pediatric reference ranges. Clin Chim Acta 2003;331:135–46.
	5	Chan MK, Seiden-Long I, Aytekin M, et al. Canadian Laboratory Initiative on Reference Interval Database (CALIPER): pediatric reference intervals for an integrated clinical chemistry and immunoassay analyzer, Abbott ARCHITECT ci8200. Clin Biochem 2009;42:885–91.
Method(s)	1	Hitachi 747 using ammonium molybdate method (Boehringer Mannheim, Diagnostics, Indianapolis, IN).
	2,3	Vitros 700 (2) and 500 (3) using ammonium molybdate method (Ortho-Clinical Diagnostics, Raritan, NJ).
	4	The phosphorus method uses a mixture of p-methylaminophenol sulfate and bisulfite to reduce the phosphomolybdate. Siemens Dimension RxL Analyzer (Siemens Healthcare Diagnostics, Deerfield, IL).
	5	Abbott Architect ci8200 (Abbott Diagnostics, Abbott Park, IL).
Comment(s)	1,3	Study used hospitalized patients and a computerized approach to removing outliers. Values are 2.5–97.5th percentiles.
	2	Study used normal healthy children. Values are 2.5–97.5th percentiles.
	4	*Numbers not provided. Reference ranges were obtained by comparing results from previously published data and using regression equations.
	5	1459 serum/plasma specimens from children attending select outpatient clinics were collected from the five age groups noted above. Values are 2.5–97.5th percentiles.

POTASSIUM

Test	Age	n	mmol/L
	Male and Female		
1	0–1 wk	100	3.2–5.5
	1 wk–1 mo	100	3.4–6.0
	1–6 mo	100	3.5–5.6
	6 mo–1 y	100	3.5–6.1
	>1 y	105	3.3–4.6
2	1–15 y	*	3.7–5.0
	16 y–Adult		3.7–4.8
3	0–1 mo	207	2.5–5.4
	1–6 mo	96	2.7–5.2
4	0–1 wk	**	3.2–5.7
	1 wk–1 mo		3.4–6.2
	1–6 mo		3.5–5.8
	6 mo–1 y		3.5–6.3
	>1 y		3.3–4.7

Specimen Type(s)	1,2	Plasma
	3	Whole blood
	4	Plasma/serum
Reference(s)	1	Greeley C, Snell J, Colaco A, et al. Pediatric reference ranges for electrolytes and creatinine. Clin Chem 1993;39:1172. (Abstract)
	2	Burritt MF, Slockbower JM, Forsman BS, et al. Pediatric reference intervals for 19 biologic variables in healthy children. Mayo Clinic Proceedings 1990;65:329–36.
	3	Snell J, Greeley C, Colaco A, et al. Pediatric reference ranges for arterial pH, whole blood electrolytes, and glucose. Clin Chem 1993;39:1173. (Abstract)
	4	Ghoshal AK, Soldin SJ. Evaluation of the Dade Behring Dimension RxL: integrated chemistry system-pediatric reference ranges. Clin Chim Acta 2003;331:135–46.
Method(s)	1	Vitros 500 (Ortho-Clinical Diagnostics, Raritan, NJ).
	2	Flame Photometry–American Monitor (American Diagnostics, Inc., Indianapolis, IN).
	3	288 Blood Gas System (Ciba Corning Diagnostics, East Walpole, MA).
	4	The sodium, potassium, and chloride (Na/K/Cl) methods use indirect sample sensing with the QuikLYTE® Integrated Multisensor Technology (IMT) to develop an electrical potential proportional to the activity of each specific ion in the sample. The total carbon dioxide (TCO_2) method uses a Severinghaus electrode designed to measure the liberated CO_2 from an acidified sample. Siemens Dimension RxL Analyzer (Siemens Healthcare Diagnostics, Deerfield, IL).
Comment(s)	1	Study used hospitalized patients and a computerized approach to removing outliers. Above values are 2.5–97.5th percentiles.
	2	From normal healthy children. *See reference for numbers.
	3	Study used hospitalized patients and a computerized approach to removing outliers. Above values are 2.5–97.5th percentiles.
	4	**Numbers not provided. Reference ranges were obtained by comparing results from previously published data and using regression equations. Above values are 2.5–97.5th percentiles.

PREALBUMIN (TRANSTHYRETIN)

		Male and Female		
Test	Age	n	mg/L	mg/dL
1	0–1 mo	63	70–390	7.0–39.0
	1–6 mo	55	83–340	8.3–34.0
	6 mo–4 y	159	20–360	2.0–36.0
	4–6 y	59	120–300	12.0–30.0
	6–19 y	189	120–420	12.0–42.0
2	0–5 d	69	60–210	6.0–21.0
	1–5 y	68	140–300	14.0–30.0
	6–9 y	68	150–330	15.0–33.0
	10–13 y	61	200–360	20.0–36.0
	14–19 y	70	220–450	22.0–45.0
3	0–4 d	118	73–144	7.3–14.4
	1 mo–4 y	116	67–171	6.7–17.1
	5–11 y	149	91–220	9.1–22.0
	12–20 y	207	124–302	12.4–30.2
4	0–5 d	*	86–232	8.6–23.2
	6 d–24 mo		96–320	9.6–32.0
	2–5 y		164–320	16.4–32.0
	6–9 y		173–349	17.3–34.9
	10–13 y		222–378	22.2–37.8
	14–19 y		242–466	24.2–46.6
5	0–12 mo	92	0–252	0–25.2
	1–5 y	146	87–266	8.7–26.6
	6–10 y	126	81–278	8.1–27.8
	11–14 y	159	142–326	14.2–32.6
	15–20 y	147	0–351	0–35.1

		Male			Female		
Test	Age	n	mg/L	mg/dL	n	mg/L	mg/dL
6	1–40 d	156	32–159	3.2–15.9	104	42–144	4.2–14.4
	41–90 d	134	27–176	2.7–17.6	90	25–219	2.5–21.9
	3–9 mo	114	73–279	7.3–27.9	94	53–250	5.3–25.0
	10–24 mo	106	67–285	6.7–28.5	80	73–337	7.3–33.7
	2–10 y	325	69–312	6.9–31.2	138	80–352	8.0–35.2
	11–15 y	153	63–335	6.3–33.5	169	86–407	8.6–40.7
	16–18 y	67	80–416	8.0–41.6	89	137–441	13.7–44.1

Specimen Type(s)	1–3,6	Serum
	4,5	Plasma/serum
Reference(s)	1	Davis ML, Austin C, Messmer BL, et al. IFCC-standardized pediatric reference intervals for 10 serum proteins using the Beckman Array 360 system. Clin Biochem 1996;29:489–92.
	2	Lockitch G, Halstead AZ, Quigley G, et al. Age- and sex-specific pediatric reference intervals: study design and methods illustrated by measurement of serum proteins with the Behring LN nephelometer. Clin Chem 1988;34:1618–21.
	3	Hamlin CR, Pankowsky DA. Turbidimetric determination of transthyretin (prealbumin) with a centrifugal analyzer. Clin Chem 1987;33:144–6.
	4	Ghoshal AK, Soldin SJ. Evaluation of the Dade Behring Dimension RxL: integrated chemistry system-pediatric reference ranges. Clin Chim Acta 2003;331:135–46.
	5	Chan MK, Seiden-Long I, Aytekin M, et al. Canadian Laboratory Initiative on Reference Interval Database (CALIPER): pediatric reference intervals for an integrated clinical chemistry and immunoassay analyzer, Abbott ARCHITECT ci8200. Clin Biochem 2009;42:885–91.
	6	Soldin S, Zhang M, Schaub JR, Ghoshal A. Pediatric reference ranges for prealbumin and retinol binding protein. Clin Biochem 2003;36:486. (Abstract)
Method(s)	1	Rate nephelometry. Beckman Array 360 (Beckman Instruments, Brea, CA).
	2	Nephelometry using the Behring LN nephelometer (Behring Diagnostics, Westwood, MA).
	3	Immunoturbidimetric procedure using a Cobas-Bio centrifugal analyzer (Roche Analytical Instruments, Inc., Nutley, NJ) and rabbit antiserum (Behring Diagnostics, La Jolla, CA).
	4	The prealbumin method is based on a particle-enhanced turbidimetric immunoassay (PETIA) technique. Siemens Dimension RxL Analyzer (Siemens Healthcare Diagnostics, Deerfield, IL).
	5	Abbott Architect ci8200 (Abbott Diagnostics, Abbott Park, IL).
	6	Nephelometry using the Behring Nephelometer 100 and Behring reagent test kits (Behring Diagnostics, Westwood, MA).
Comment(s)	1	Samples were obtained from children attending outpatient clinics. Results are 2.5–97.5th percentiles.
	2	No significant differences were found for males and females. These ranges were therefore derived from combined data. The study used healthy children. Results are 2.5–97.5th percentiles.
	3	Healthy children. Values are 2.5–97.5th percentiles.
	4	*Numbers not provided Reference ranges were obtained by comparing results from previously published data and using regression equations.
	5	1459 serum/plasma specimens from children attending select outpatient clinics were collected from the five age groups noted above. Values are 2.5–97.5th percentiles.
	6	Study used hospitalized patients in a computerized approach to removing outliers.

PREGNENOLONE

Male and Female

Age	n	ng/dL	nmol/L
Premature infants 26–28 wks, day 4	*	260–2104	8.2–66.4
Newborn 1–7 d		150–2000**	4.7–63.1**
Infants and prepubertal children		20–140	0.6–4.4
Pubertal age groups (11–16 y)		10–150	0.3–4.7

Specimen Type(s)	Serum
Reference(s)	Endocrinology expected values. Esoterix Endocrinology, Calabasas Hills, CA. ©2002 Esoterix, Inc., http://www.esoterix.com
Method(s)	Not provided.
Comment(s)	*Numbers not provided. **Levels decrease after birth, and are within the prepubertal range by three months.

17-OH-PREGNENOLONE

Male and Female

Age	n	ng/dL	nmol/L
Premature infants 26–28 wks, day 4	*	375–3559	11.3–106.9
31–35 wks, day 4		64–2380	1.9–71.5
Full-term infants 3 d		10–829	0.3–24.9
1–6 mo		36–763	1.1–22.9
6–12 mo		42–540	1.3–16.2
Prepubertal children 1–2 y		14–207	0.4–6.2
2–6 y		10–103	0.3–3.1
6–10 y		10–186	0.3–5.6
Pubertal age groups (11–16 y)		44–235	1.3–7.1

Specimen Type(s)	Serum
Reference(s)	Endocrinology expected values. Esoterix Endocrinology, Calabasas Hills, CA. ©2002 Esoterix, Inc., http://www.esoterix.com
Method(s)	Not provided.
Comment(s)	*Numbers not provided.

PROGESTERONE

Test	Age	Male			Female		
		n	ng/dL	nmol/L	n	ng/dL	nmol/L
1	6–9 y	23*	≤20	≤0.64	23*	≤20	≤0.64
	10–11 y	25*	≤20	≤0.64	25*	≤20	≤0.64
	12–17 y	24	≤20	≤0.64			

Test	Age	Male and Female		
		n	ng/dL	nmol/L
2	Premature infants	**		
	26–28 wks		18–640	0.57–20.35
	31–35 wks		84–1360	2.67–43.25
	Full-term infants			
	1–7 d		a	a
	Prepubertal children			
	1–10 y		7–52	0.22–1.65

Male			Female[b]		
Age	ng/dL	nmol/L	Age	ng/dL	nmol/L
Puberty[c]			Puberty[c]		
Tanner Stage[d]			Tanner Stage[d]		
1 (<9.8 y)	<10–33	<0.32–1.06	1 (<9.2 y)	<10–33	<0.32–1.06
2 (9.8–14.5 y)	<10–33	<0.32–1.06	2 (9.2–13.7 y)	<10–55	<0.32–1.76
3 (10.7–15.4 y)	<10–48	<0.32–1.54	3 (10.0–14.4 y)	10–450	0.32–14.4
4 (11.8–16.2 y)	10–108	0.32–3.46	4 (10.7–15.6 y)	10–1300	0.32–41.6
5 (12.8–17.3 y)	21–82	0.67–2.62	5 (11.8–18.6 y)	10–950	0.32–30.4
n**			n**		

Specimen Type(s)	1,2	Serum
Reference(s)	1	Pediatric endocrine testing. Nichols Institute, 1993:34.
	2	Endocrinology expected values. Esoterix Endocrinology, Calabasas Hills, CA. ©2002 Esoterix, Inc., http://www.esoterix.com
		[c]Data adapted from Bidlingmaier F, Wagner-Barnack M, Butenandt O, Knorr D. Plasma estrogens in childhood and puberty under physiologic and pathologic conditions. Pediat Res 1973;7:901.
		[d]Tanner JM, Whitehouse RH. Clinical longitudinal standards for height, weight, height velocity, and the stages of puberty. Arch Dis Childhood 1976;51:170.
Method(s)	1	Extraction, radioimmunoassay.
	2	Not provided.
Comment(s)	1	Results are 2.5–97.5th percentiles.
		*Males and females combined
	2	**Numbers not provided.
		[a]Progesterone levels are markedly elevated in the neonate, but fall rapidly to reach prepubertal levels by seven days, where they remain until puberty.
		[b]Day of cycle not determined for prepubertal females.
		[d]As used herein, Tanner stages in males encompass development of both pubic hair and genitalia. In females, each stage encompasses both pubic hair and breast development. While this expands the chronological age range for each stage of pubertal development, it results in a better correlation with hormonal values.

PROLACTIN

Test	Age	Male		Female	
		n	ng/mL (µg/L)	n	ng/mL (µg/L)
1	0–1 mo	45	3.7–81.2	43	0.3–95.0
	1–12 mo	99	0.3–28.9	89	0.2–29.9
	1–3 mo	141	2.3–13.2	114	1.0–17.1
	4–6 y	106	0.8–16.9	139	1.6–13.1
	7–9 y	105	1.9–11.6	98	0.3–12.9
	10–12 y	109	0.9–12.9	108	1.9–9.6
	13–15 y	138	1.6–16.6	122	3.0–14.4
	16–18 y	95	2.7–15.2	104	2.1–18.4
2	<2 y	158	2.7–25.0	41	4.2–20.2
	2–5 y	142	1.7–17.0	91	1.0–17.5
	6–10 y	103	0.7–15.8	140	1.2–15.5
	11–20 y	91	2.0–18.2	280	1.5–19.5
3	<2 y	158	2.3–18.3	41	3.3–14.7
	2–5 y	142	1.6–12.5	91	1.0–12.8
	6–10 y	103	0.8–11.6	140	1.2–11.4
	11–20 y	91	1.8–13.3	280	1.4–14.3

		Male and Female	
Test	Age	n	ng/mL (µg/L)
4	0–12 mo	86	3.3–109.7
	1–5 y	141	3.9–31.5
	6–10 y	126	2.9–35.0

Age	Male		Female	
	n	ng/mL (µg/L)	n	ng/mL (µg/L)
11–14 y	67	3.5–16.3	92	4.6–49.1
15–20 y	37	4.6–21.0	90	4.2–117.9

Specimen Type(s)	1–4	Plasma/serum
Reference(s)	1	Cook JF, Hicks JM, Godwin ID, et al. Pediatric reference ranges for prolactin. Clin Chem 1992;38:959. (Abstract)
	2	Soldin SJ, Morales A, Albalos F, et al. Pediatric reference ranges on the Abbott IMx analyzer for FSH, LH, prolactin TSH, T_4, T_3, free T_4, free T_3, T uptake, IgE and ferritin. Clin Biochem 1995;28:603–6.
	3	Murthy JN, Hicks JM, Soldin SJ. Evaluation of the Technicon Immuno I Random Access Immunoassay Analyzer and calculation of pediatric reference ranges for endocrine tests, T-uptake, and ferritin. Clin Biochem 1995;28:181–5.
	4	Chan MK, Seiden-Long I, Aytekin M, et al. Canadian Laboratory Initiative on Reference Interval Database (CALIPER): pediatric reference intervals for an integrated clinical chemistry and immunoassay analyzer, Abbott ARCHITECT ci8200. Clin Biochem 2009;42:885–91.
Method(s)	1	Hybritech Tandem Prolactin procedure (Hybritech Inc, San Diego, CA).
	2	IMx Analyzer (Abbott Laboratories, Abbott Park, IL).
	3	Immuno I (Bayer Corp., Tarrytown, NY).
	4	Abbott Architect ci8200 (Abbott Diagnostics, Abbott Park, IL).
Comment(s)	1–3	Studies used hospitalized patients and a computerized approach adapted from the Hoffmann technique. Values are 2.5–97.5th percentiles.
	4	1459 serum/plasma specimens from children attending select outpatient clinics were collected from the five age groups noted above. Values are 2.5–97.5th percentiles.

PROTEIN, TOTAL

Test	Age	\multicolumn{3}{c}{Male}			\multicolumn{3}{c}{Female}		
		n	g/dL	g/L	n	g/dL	g/L
1	1–60 d	62	3.9–7.6	39–76	69	3.4–7.0	34–70
	61–180 d	130	4.1–7.9	41–79	161	3.9–7.6	39–76
	181 d–1 y	268	3.9–7.9	39–79	196	4.5–7.8	45–78
2	0–5 d (<2.5 kg)	30	3.8–6.2	38–62	30	3.8–6.2	38–62
	0–5 d (>2.5 kg)	93	5.4–7.0	54–70	93	5.4–7.0	54–70
	1–3 y	50	5.9–7.0	59–70	50	5.9–7.0	59–70
	4–6 y	38	5.9–7.8	59–78	38	5.9–7.8	59–78
	7–9 y	74	6.2–8.1	62–81	74	6.2–8.1	62–81
	10–19 y	332	6.3–8.6	63–86	332	6.3–8.6	63–86
3	1–30 d	68	4.1–6.3	41–63	51	4.2–6.2	42–62
	31–182 d	58	4.7–6.7	47–67	42	4.4–6.6	44–66
	183–365 d	29	5.5–7.0	55–70	186	5.6–7.9	56–79
	1–18 y	652	5.7–8.0	57–80	440	5.7–8.0	57–80
4	1–60 d	*	4.0–7.6	40–76	*	3.6–7.0	36–70
	61–180 d		4.0–7.0	40–70		4.0–7.6	40–76
	181 d–1 y		4.2–7.9	42–79		4.6–7.8	46–78
	1–6 y		6.0–8.0	60–80		6.0–7.8	60–78
	7–9 y		6.3–8.1	63–81		6.3–8.1	63–81
	10–19 y		6.4–8.6	64–86		6.4–8.6	64–86

Test	Age	\multicolumn{3}{c}{Male and Female}		
		n	g/dL	g/L
5	0–12 mo	129	4.9–7.3	49–73
	1–5 y	155	6.2–8.0	62–80
	6–10 y	135	6.6–8.6	66–86
	11–14 y	154	6.4–8.5	64–85
	15–20 y	180	6.4–8.3	64–83

Specimen Type(s)	1,3,4,5	Plasma/serum
	2	Plasma
Reference(s)	1	Soldin SJ, Morse AS. Pediatric reference ranges for albumin and total protein in children <1 year old using the Vitros 500 analyzer. Clin Chem 1998; 44:A15. (Abstract)
	2	Lockitch G, Halstead AC, Albersheim S, et al. Age- and sex-specific pediatric reference intervals for biochemistry analytes as measured with the Ektachem-700 analyzer. Clin Chem 1988;34:1622–5.
	3	Hicks JM, Bjorn S, Beatey J, et al. Pediatric reference ranges for albumin, globulin and total protein on the Hitachi 747. Clin Chem 1995;41:S93. (Abstract)
	4	Ghoshal AK, Soldin SJ. Evaluation of the Dade Behring Dimension RxL: integrated chemistry system-pediatric reference ranges. Clin Chim Acta 2003;331:135–46.
	5	Chan MK, Seiden-Long I, Aytekin M, et al. Canadian Laboratory Initiative on Reference Interval Database (CALIPER): pediatric reference intervals for an integrated clinical chemistry and immunoassay analyzer, Abbott ARCHITECT ci8200. Clin Biochem 2009;42:885–91.
Method(s)	1	Vitros 500 (Ortho-Clinical Diagnostics, Raritan, NJ).
	2	Biuret method. Vitros 700 (Ortho-Clinical Diagnostics, Raritan, NJ).
	3	Boehringer Mannheim total protein reagent (Biuret). Total protein was measured on the Hitachi 747 (Boehringer Mannheim Diagnostics, Indianapolis, IN).
	4	The total protein method is a modification of the biuret reaction. Siemens Dimension RxL Analyzer (Siemens Healthcare Diagnostics, Deerfield, IL).
	5	Abbott Architect ci8200 (Abbott Diagnostics, Abbott Park, IL).
Comment(s)	1,3	Study used hospitalized patients and a computerized approach to removing outliers. Values are 2.5–97.5th percentiles.
	2	Healthy term and pre-term neonates and normal healthy children. Values are 2.5–97.5th percentiles.
	4	*Numbers not provided. Reference ranges were obtained by comparing results from previously published data and using regression equations.
	5	1459 serum/plasma specimens from children attending select outpatient clinics were collected from the five age groups noted above. Values are 2.5–97.5th percentiles.

PROTEIN CEREBROSPINAL FLUID (CSF PROTEIN)

Test	Age	Male n	Male mg/dL	Male mg/L	Female n	Female mg/dL	Female mg/L
1	0–14 d	74	<117	<1170	49	<184	<1840
	15–30 d	92	<112	<1120	70	<118	<1180
	31–90 d	223	<88	<880	150	<108	<1080
	3–6 mo	54	<52	<520	53	<46	<460
	7–24 mo	98	<52	<520	77	<54	<540
	2–7 y	101	<54	<540	142	<51	<510
	8–12 y	89	<48	<480	104	<48	<480
	13–18 y	76	<42	<420	85	<48	<480
2	0–14 d	*	15–100	150–1000	*	15–153	150–1530
	15–30 d		15–96	150–960		15–100	150–1000
	31–90 d		15–48	150–480		15–93	150–930
	3–6 mo		15–48	150–480		15–44	150–440
	7–24 mo		15–50	150–500		15–48	150–480
	2–7 y		15–45	15–450		15–45	150–450
	8–12 y		15–40	15–400		15–45	150–450
	13–18 y		15–40	15–400		15–45	150–450

Specimen Type(s)	1,2	Cerebrospinal fluid
Reference(s)	1	Soldin SJ, Baumel CR. Pediatric reference ranges for CSF protein and plasma lactate on the Vitros 500 analyzer. Clin Chem 2001;47:A108. (Abstract)
	2	Ghoshal AK, Soldin SJ. Evaluation of the Dade Behring Dimension RxL: integrated chemistry system-pediatric reference ranges. Clin Chim Acta 2003;331:135–46.
Method(s)	1	Vitros 500 analyzer using Ortho-Clinical Diagnostics reagents (Ortho-Clinical Diagnostics, Raritan, NJ).
	2	Pyrogallol red combines with sodium molybdate to form a red complex with maximum absorbance at 470 nm. The protein in the sample reacts with this complex in acid solution to form a bluish-purple colored complex, which absorbs at 600 nm. Siemens Dimension RxL Analyzer (Siemens Healthcare Diagnostics, Deerfield, IL).
Comment(s)	1	Studies used hospitalized patients and a computerized approach adapted from the Hoffmann technique. Values are 97.5th percentile.
	2	*Numbers not provided. Reference ranges were obtained by comparing results from previously published data and using regression equations.

PYRUVATE

Male and Female

Test	Age	n	mg/dL	μmol/L
1	All ages	*	0.70–1.32	80–150
2	All ages	*	0.3–0.70	35–80

Specimen Type(s)	1,2	Whole blood, proteins precipitated with perchloric acid.
Reference(s)	1	Reference values and SI unit information. The Hospital for Sick Children, Toronto, 1993:378.
	2	Children's National Medical Center, Washington, DC. Unpublished data.
Method(s)	1,2	Standard enzymatic approach.
Comment(s)	1,2	*Numbers not available.

RENIN (PLASMA)

Male and Female

Age	n	ng/L
1–6 y	63	6.3–149
7–12 y	68	5.5–110
13–17 y	87	3.3–61

Specimen Type(s)	Plasma
Reference(s)	Coates JE, Chapelski LJ, Yatscoff RW. Pediatric reference intervals for plasma renin. Clin Biochem 1994;27:316–7.
Method(s)	Renin Active Pasteur CT kits (Sanofi Diagnostics Pasteur, Inc., Montreal, Quebec, Canada).
Comment(s)	Children seen as outpatients were tested. Male and female ranges were not significantly different and they were combined for calculations of the above ranges. The results reported are 2.5–97.5th percentiles.

RENIN ACTIVITY

		Male and Female
Age	n	ng/mL/h
0–1 wk	7	0–40
2–4 wk	7	0–175
3 mo–1 y	*	≤15
1–4 y	*	≤10
4–15 y	*	≤6

Specimen Type(s)	Plasma
Reference(s)	Pediatric endocrine testing. Nichols Institute, 1993:33.
Method(s)	Radioimmunoassay.
Comment(s)	Normal sodium diet. Patients over 1 y of age were supine from 9–11 am before specimens were drawn. *Numbers not provided.

RETINOL BINDING PROTEIN (RBP)

Test	Age	n	mg/dL	µmol/L
	Male and Female			
1	0–5 d	64	0.8–4.5	0.38–2.15
	1–5 y	68	1.0–7.6	0.48–3.62
	6–9 y	64	2.0–7.8	0.95–3.72
	10–13 y	59	1.3–9.9	0.62–4.72
	14–19 y	70	3.0–9.2	1.43–4.39
2	Term neonates (>37 wk)	41	1.02–3.14	0.49–1.51
	Preterm neonates (<36.6 wk)	58	0.72–2.88	0.35–1.38

Test	Age	Male			Female		
		n	mg/dL	µmol/L	n	mg/dL	µmol/L
3	0–30 d	114	0.91–5.6	0.43–2.67	92	1.01–5.75	0.48–2.74
	1 mo–18 y	96	1.09–8.7	0.52–4.15	97	0.98–6.76	0.47–3.22

Specimen Type(s)	1	Serum
	2,3	Plasma/serum
Reference(s)	1	Lockitch G, Halstead AZ, Quigley G, et al. Age- and sex-specific pediatric reference intervals: study design and methods illustrated by measurement of serum proteins with the Behring LN nephelometer. Clin Chem 1988;34:1618–21.
	2	Cardona-Pérez A, et al. Cord blood retinol and retinol binding protein in preterm and term neonates. Nutrition Research 1996;16:191–6.
	3	Soldin S, Zhang M, Schaub JR, Ghoshal A. Pediatric reference ranges for prealbumin and retinol binding protein. Clin Biochem 2003;36:486. (Abstract)
Method(s)	1,3	Nephelometry using the Behring LN nephelometer (Behring Diagnostics, Westwood, MA).
	2	Radial Immunodiffusion (Behring Diagnostics, La Jolla, CA).
Comment(s)	1	No significant differences were found for males and females. These ranges were therefore derived from combined data. The study used healthy children. Results are 2.5–97.5th percentiles.
	2	Healthy term and preterm neonates studied. Results are 2.5–97.5th percentiles.
	3	Study used hospitalized patients in a computerized approach to removing outliers.

REVERSE TRIIODOTHYRONINE (rT$_3$)

		Male and Female	
Age	n	ng/dL	nmol/L
1–5 y	*	15–71	0.23–1.10
5–10 y		17–79	0.26–1.20
10–15 y		19–88	0.29–1.36
Adults		30–80	0.46–1.23
Specimen Type(s)	Serum		
Reference(s)	Behrman RE, ed. Nelson textbook of pediatrics, 14th ed. Philadelphia, PA: WB Saunders Company, 1992:1821.		
Method(s)	Not given.		
Comment(s)	*Numbers not provided.		

SELENIUM

		Male and Female		
Test	Age	n	µg/dL	µg/dL
1	0–5 d*	20	6–9	0.72–1.20
	1–5 y*	30	10–14	1.22–1.82
	6–9 y*	30	10–16	1.29–2.05
	10–14 y*	30	10–19	1.31–2.35
	15–19 y*	27	10–19	1.31–2.35
2	3–5 y	38	2.0–12.4	0.25–1.57
	6–8 y	43	1.6–12.0	0.20–1.52
	9–11 y	52	2.4–13.5	0.30–1.70
	12–13 y	24	4.0–10.7	0.51–1.35
	14–16 y	29	4.0–10.7	0.51–1.35
3	0–<0.5 y	13	0.8–4.4	0.10–0.58
	0.5–<1 y	18	0–6.9	0.00–0.87
	1–<2 y	15	1.5–7.3	0.20–0.96
	2–<4 y	23	0.9–10.4	0.12–1.36
	4–<6 y	19	0.6–12.5	0.08–1.64
	6–<10 y	25	2.4–11.8	0.31–1.55
	10–<14 y	21	2.4–11.4	0.30–1.50
	14–<18 y	17	2.1–11.8	0.27–1.55

Specimen Type(s)	1,2	Serum
	3	Plasma/serum
Reference(s)	1	Lockitch G, Halstead A, Wadsworth L, et al. Age- and sex-specific pediatric reference intervals for zinc, copper, selenium, iron, vitamins A and E, and related proteins. Clin Chem 1988;34:1625–8.
	2	Malvy DJ–M, Arnaud J, Burtschy B, et al. Reference values for serum, zinc and selenium of French healthy children. Eur J Epidemiol 1993;9:155–61.
	3	Rükgauer M, Klein J, Kruse-Jarres JD. Reference values for the trace elements copper, manganese, selenium, and sinc in the serum/plasma of children, adolescents, and adults. J Trace Elements Med Biol 1997;11:92–8.
Method(s)	1	Varian GTA-95. Atomic absorption (Varian Canada, Inc., Georgetown, Ontario, Canada).
	2	Perkin-Elmer atomic absorption spectrophotometry (Perkin Elmer Corporation, Norwalk, CT).
	3	Atomic absorption spectrophotometry with Zeeman background compensation. Perkin Elmer ETAAS, Zeeman 3030 (Uberlingen, Germany).
Comment(s)	1	The study population was healthy children. Non-parametric methods were used to determine the 0.025 and 0.975 fractiles. *No significant differences were found for males and females. These ranges were therefore derived from combined data.
	2	The study population consisted of healthy French children. Values reported are the 2.5–97.5th percentiles. *Note:* Reference ranges differ greatly on a geographic basis. Data must be interpreted against locally derived reference ranges.
	3	Study population was drawn from patients visiting the outpatient department or surgical or orthopedic ward for preoperative work up. Results are mean ±2 SDs (2.5–97.5th percentiles).

SERUM OSTEOCALCIN

Age	Male		Female	
	n	µg/L	n	µg/L
1 y	43	52–63	31	48–79
2 y	24	64–70	31	61–88
3 y	39	54–70	35	44–93
4 y	43	53–90	47	51–98
5 y	55	52–83	49	66–88
6 y	59	51–84	66	65–96
7 y	51	49–104	42	53–97
8 y	66	51–80	78	50–89
9 y	70	62–83	57	61–78
10 y	86	58–88	82	65–96
11 y	78	56–80	69	36–93
12 y	41	68–80	39	74–99
13 y	47	74–98	35	42–75
14 y	55	66–102	51	25–43
15 y	49	61–105	50	35–59
16 y	27	36–50	39	27–47

Specimen Type(s)	Serum
Reference(s)	Cioffi M, Molinari AM, Gazzero P, et al. Serum osteocalcin in 1634 healthy children. Clin Chem 1997;43:543–5.
Method(s)	Sandwich IRMA (Osteo-ELSA; CIS Bio International, Gif sur Yvette, France).
Comment(s)	Study used 1634 healthy children. Results are 25th and 75th percentiles. Osteocalcin is an important marker of bone turnover.

SEX HORMONE BINDING GLOBULIN (SHBG)

Age	Male			Female		
	n	nmol/L	µg/dL	n	nmol/L	µg/dL
1–30 d	57	10.8–70.8	0.31–2.04	49	11.8–51.4	0.34–1.48
31–365 d	93	60.2–208.5	1.73–6.01	68	50.4–181.2	1.45–5.22
1–3 y	109	42.4–155.6	1.22–4.48	98	51.4–157.7	1.48–4.54
4–6 y	75	39.4–145.6	1.14–4.20	73	47.8–142.1	1.38–4.10
7–9 y	70	37.7–114.4	1.09–3.30	29	31.0–103.0	0.89–2.97
10–12 y	113	31.6–92.5	0.91–2.67	60	20.0–99.6	0.58–2.87
13–15 y	78	13.3–62.6	0.38–1.80	80	16.6–76.5	0.48–2.20
16–18 y	34	10.6–53.6	0.31–1.54	39	9.3–75.2	0.27–2.17

Specimen Type(s)	Plasma/serum
Reference(s)	Soldin SJ, Hicks JM, Bailey J, et al. Pediatric reference ranges for sex hormone binding globulin. Clin Chem 1997;43:S200. (Abstract)
Method(s)	Time-resolved fluoroimmunoassay. Delfia SHBG kit (Wallac Oy, Turku, Finland).
Comment(s)	Study used hospitalized patients and a computerized adaptation of the Hoffmann technique. Values are 2.5–97.5th percentiles.

SODIUM

		Male and Female	
Test	Age	n	mmol/L
1	0–7 d	100	133–146
	7–31 d	100	134–144
	1–6 mo	100	134–142
	6 mo–1 y	100	133–142
	>1 y	105	134–143
2	0–1 mo	59	127–143
	2–6 mo	49	130–147
3	0–7 d	*	131–144
	7–31 d		132–142
	1–6 mo		132–140
	6 mo–1 y		131–140
	>1 y		132–141

Specimen Type(s)	1	Plasma
	2	Heparinized whole blood
	3	Plasma/serum
Reference(s)	1	Greeley C, Snell J, Colaco A, et al. Pediatric reference ranges for electrolytes and creatinine. Clin Chem 1993;39:1172. (Abstract)
	2	Snell J, Greeley C, Colaco A, et al. Pediatric reference ranges for arterial pH, whole blood electrolytes and glucose. Clin Chem 1993;39:1173. (Abstract)
	3	Ghoshal AK, Soldin SJ. Evaluation of the Dade Behring Dimension RxL: integrated chemistry system-pediatric reference ranges. Clin Chim Acta 2003;331:135–46.
Method(s)	1	Vitros 700 analyzer (Ortho-Clinical Diagnostics, Raritan, NJ).
	2	Corning 288 Blood Gas System (Ciba Corning Diagnostics, East Walpole, MA).
	3	The sodium, potassium, and chloride (Na/K/Cl) methods use indirect sample sensing with the QuikLYTE® Integrated Multisensor Technology (IMT) to develop an electrical potential proportional to the activity of each specific ion in the sample. Siemens Dimension RxL Analyzer (Siemens Healthcare Diagnostics, Deerfield, IL).
Comment(s)	1,2	Study used hospitalized patients and a computerized approach adapted from the Hoffmann technique to obtain the 2.5–97.5th percentiles.
	3	*Numbers not provided. Reference ranges were obtained by comparing results from previously published data and using regression equations.

SWEAT ELECTROLYTES

Test	Age	n	mmol/L	mmol/L
	Male and Female			
			Chloride	Sodium
1	1 wk–adult	*	<40	<40
	Patients with cystic fibrosis		>60	>60
2	0–3.9 mo	170	6–40	4–40
	4–11.9 mo	70	4–29	5–39
	1–4.9 y	134	6–54	6–54

Specimen Type(s)	1,2	Sweat
Reference(s)	1	Children's National Medical Center, Washington, DC. Unpublished data.
	2	Meites S, ed. Pediatric clinical chemistry, 3rd ed. Washington, DC: AACC Press, 1989:246.
Method(s)	1	Pilocarpine iontophoresis/macroduct collection. Measured on Vitros 500 (Ortho-Clinical Diagnostics, Raritan, NJ). Cl measured on Chloridometer (Radiometer, Cleveland, OH).
	2	Pilocarpine iontophoresis. Flame photometry for sodium, chloridometer for chloride.
Comment(s)	1	Results between 40–60 mmol/L require repeat testing and/or use of other tests such as immunoreactive trypsin, bentiromide test of pancreatic function, and DNA probe test for CF gene.
		*Numbers not provided.
	2	Based on in-house studies of non-affected individuals.

TESTOSTERONE

Test	Age	Male			Female		
		n	ng/dL	nmol/L	n	ng/dL	nmol/L
1	1–5 mo	10	1–177	0.03–6.14	5	1–5	0.03–0.17
	6–11 mo	8	2–7	0.07–0.24	6	2–5	0.07–0.17
	1–5 y	16	2–25	0.07–0.87	17	2–10	0.07–0.35
	6–9 y	30	3–30	0.10–1.04	16	2–20	0.07–0.35
	10–11 y	20	5–50	0.17–1.73	10	5–25	0.17–0.87
	12–14 y	28	10–572	0.35–19.83	17	10–40	0.35–1.39
	15–17 y	18	220–800	7.63–27.74	11	5–40	0.17–1.39
2	Premature infants	*			*		
	26–28 wks, day 4		59–125	2.1–4.4		5–16	0.18–0.6
	31–35 wks, day 4		37–198	1.3–6.9		5–22	0.18–0.8
	Full-term infants						
	Newborns		75–400	2.6–14		20–64	0.7–2.2
	1–7 mo[a,b]		c	c		d	d
	Prepubertal children						
	1–10 y		<3–10	<0.1–0.4		<3–10	<0.11–0.4

Male			Female[e]		
Age	ng/dL	nmol/L	Age	ng/dL	nmol/L
Puberty[f]			Puberty[f]		
Tanner Stage[g]			Tanner Stage[g]		
1 (<9.8 y)	<3–10	<0.1–0.4	1 (<9.2 y)	<3–10	<0.1–0.4
2 (9.8–14.5 y)	18–150	0.6–5.3	2 (9.2–13.7 y)	7–28	0.3–1.0
3 (10.7–15.4 y)	100–320	3.5–11.2	3 (10.0–14.4 y)	15–35	0.5–1.2
4 (11.8–16.2 y)	200–620	7.0–21.7	4 (10.7–15.6 y)	13–32	0.5–1.1
5 (12.8–17.3 y)	350–970	12.3–34.0	5 (11.8–18.6 y)	20–38	0.7–1.3
n*			n*		

Test	Male				Female			
	Age	n	ng/dL	nmol/L	Age	n	ng/dL	nmol/L
3	0–6 y	158	4–31	0.1–1.1	0–5 y	127	2–10	0.1–0.3
	7–9 y	148	4–25	0.1–0.9	6–9 y	171	5–13	0.2–0.5
	10–12 y	188	5–418	0.2–14.5	10–14 y	190	14–50	0.5–1.9
	13–14 y	245	6–647	0.2–22.5	15–16 y	175	12–53	0.4–1.8
	15–16 y	209	42–880	1.5–30.5	17–18 y	123	16–50	0.6–1.9
	17–19 y	123	121–823	4.2–28.5				

Test	Male				Female			
	Age	n	ng/dL	nmol/L	Age	n	ng/dL	nmol/L
4	0–<1 y	18	2–10	<0.069*–0.338	0–<1 y	29	2–20	<0.069*–0.695
	Male and Female							
	Age		n		ng/dL		nmol/L	
	>1–≤5 y		97		2.0–11.4		<0.069*–0.394	
	>5–≤10 y		96		2.0–22.2		<0.069*–0.820	
	Male				Female			
	Age	n	ng/dL	nmol/L	Age	n	ng/dL	nmol/L
	>10–≤15 y	49	2–830	<0.069*–28.797	>10–≤20 y	80	2–78	<0.069*–2.715
	>15–≤20 y	25	103–1011	3.547–35.075				

Specimen Type(s)	1–3	Serum
	4	Plasma/serum
Reference(s)	1	Pediatric endocrine testing. Nichols Institute, 1993:38.
	2	Endocrinology expected values. Esoterix Endocrinology, Calabasas Hills, CA. ©2002 Esoterix, Inc., http://www.esoterix.com
		[a]Data adapted from Pang S, Levine LS, Chow D, Sagiani F, Saenger P, New MI. Dihydrotestosterone and its relationship to testosterone in infancy and childhood. J Clin Endocrinol Metab 1979;48:821.
		[b]Data adapted from Forest MG, Cathiard AM, Bertrand JA. Evidence of testicular activity in early infancy. J Clin Endocrinol Metab 1973;37:148.
		[f]Data adapted from Bidlingmaier F, Wagner-Barnack M, Butenandt O, Knorr D. Plasma estrogrens in childhood and puberty under physiologic and pathologic conditions. Pediat Res 1973;7:901.
		[g]Tanner JM, Whitehouse RH. Clinical longitudinal standards for height, weight, height velocity, and the stages of puberty. Arch Dis Childhood 1976;51:170.
	3	Soldin OP, Sharma H, Husted L, Soldin SJ. Pediatric reference intervals for aldosterone, 17alpha-hydroxyprogesterone, dehydroepoandrosterone, testosterone and 25-hydroxy vitamin D3 using tandem mass spectrometry. Clin Biochem 2009;42:823–7.
	4	Kulasingam V, Jung BP, Blasutig IM, et al. Pediatric reference intervals for 28 chemistries and immunoassays on the Roche cobas® 6000 analyzer—A CALIPER pilot study. Clin Biochem 2010;43:1045–50.
Method(s)	1	Extraction, chromatography, radioimmunoassay.
	2	Not provided.
	3	Samples were analyzed using isotope dilution liquid chromatography tandem mass spectrometry (LC/MS/MS).
	4	Roche cobas® 6000 analyzer (Roche Diagnostics Limited, West Sussex, UK).
Comment(s)	1	Results are 2.5–97.5th percentiles.
	2	*Numbers not provided.
		[c]Levels decrease rapidly the first week to 20–50 ng/dL, then increase to 60–400 ng/dL (mean = 190) between 20–60 days. Levels then decline to prepubertal range by seven months.
		[d]Levels decrease during the first month to <10 ng/dL and remain there until puberty.
		[e]Day of cycle not determined for pubertal females.
		[g]As used herein, Tanner stages in males encompass development of both pubic hair and genitalia. In females, each stage encompasses both pubic hair and breast development. While this expands the chronological age range for each stage of pubertal development, it results in a better correlation with hormonal values.
	3	Reference intervals were determined for neonates and children 0–18 years of age. The study was conducted using outpatient samples obtained between January 1, 2004, and June 30, 2008. Values are the 2.5–97.5th percentiles.
	4	*Lower reference interval at the limit of detection of the assay.
		Approximately 600 outpatient samples from a pediatric population deemed to be metabolically stable were subdivided into five age classes ranging from 0 to 20 years of age and further partitioned by gender. Values are 2.5–97.5th percentiles.

THYROID STIMULATING HORMONE (TSH)

Test	Age	Male n	Male mIU/L	Female n	Female mIU/L
1	0–1 mo	84	<6.5	62	<6.0
	1–12 mo	114	<4.1	103	<4.0
	1–3 y	128	<3.0	126	<3.3
	4–6 y	109	<3.0	82	<2.8
	7–12 y	112	<3.1	107	<2.9
	13–18 y	106	<3.1	106	<3.0
2	1–30 d	63	0.5–16.0	89	0.7–13.1
	1 mo–5 y	95	0.6–7.1	95	0.5–8.1
	6–18 y	95	0.4–6.0	96	0.4–5.8
3	1–30 d	63	0.7–16.1	89	1.0–13.7
	1 mo–5 y	95	0.8–7.5	95	0.7–8.6
	6–18 y	95	0.6–6.4	96	0.6–6.2

Male and Female

Test	Age	n	mIU/L
4	0–<2 y	36	0.7–4.5
	2–<7 y	149	0.4–3.2
	7–<13 y	128	0.3–2.7
	13–<18 y	123	0.4–1.9
5	0–3 d	*	1.0–20.0
	3–30 d		0.5–6.5
	1–5 mo		0.5–6.0
	6 mo–18 y		0.5–4.5
6	0–12 mo	71	0.9–5.4

Test	Age	Male n	Male mIU/L	Female n	Female mIU/L
	1–5 y	152	0.7–4.5	155	0.7–4.8
	11–14 y	93	0.6–3.6	201	0.5–4.1
7	<2 mo	212	1.1–6.3	171	1.1–5.5
	2–<20 mo	138	1.0–4.9	152	1.1–4.5
	12–24 mo	197	0.9–4.8	203	1.0–4.4
	2–<5 y	297	0.8–4.4	264	0.9–4.1
	5–<10 y	537	0.8–4.1	697	0.9–4.1
	10–<15	746	0.8–4.0	1063	0.7–3.7
	15–<20 y	547	0.6–3.6	877	0.5–3.6

Test	Age	n	Male mIU/L	Female mIU/L
8	1–30 d	**	0.6–12.8	0.8–10.8
	1 mo–5 y		0.7–6.0	0.6–6.8
	6–18 y		0.5–5.1	0.5–4.9

Test	Age	n	Male and Female mIU/L		
9	0–≤5 y	189	0.8–6.2		
	Age	Male n	Male mIU/L	Female n	Female mIU/L
	>5–≤10 y	47	1.2–5.3	61	0.5–4.8
	Age	n	Male and Female mIU/L		
	>10–≤15 y	118	0.8–4.2		
	Age	Male n	Male mIU/L	Female n	Female mIU/L
	>15–≤20 y	26	0.6–5.4	35	0.4–2.8

Specimen Type(s)	1–4, 6–9	Plasma/serum
	5	Serum
Reference(s)	1	Hicks JM, Godwin ID, Beatey J, et al. Pediatric reference ranges for highly sensitive TSH. Clin Chem 1992;38:960. (Abstract)
	2	Soldin SJ, Morales A, Albalos F, et al. Pediatric reference ranges on the Abbott IMx for FSH, LH, prolactin, TSH, T_4, T_3, free T_4, free T_3, T-uptake, IgE and ferritin. Clin Biochem 1995;28:603–6.
	3	Murthy JN, Hicks JM, Soldin SJ. Evaluation of the Technicon Immuno I Random Access Immunoassay Analyzer and calculation of pediatric reference ranges for endocrine tests, T-uptake, and ferritin. Clin Biochem 1995;28:181–5.
	4	Soldin SJ, Hicks JM, Bailey J, et al. Pediatric reference ranges for 3rd generation TSH on the ACS 180. Clin Chem 1998; 44:A13. (Abstract)
	5	Dugaw KA, Jack RM, Rutledge J. Pediatric reference ranges for TSH, free T_4, total T_4, total T_3 and T_3 uptake on the Vitros ECi analyzer. Chem Chem 2001;47:A108. (Abstract)
	6	Chan MK, Seiden-Long I, Aytekin M, et al. Canadian Laboratory Initiative on Reference Interval Database (CALIPER): pediatric reference intervals for an integrated clinical chemistry and immunoassay analyzer, Abbott ARCHITECT ci8200. Clin Biochem 2009;42:885–91.
	7	Soldin SJ, Cheng LL, Lam LY, et al. Comparison of FT4 with log TSH on the Abbott Architect ci8200: pediatric reference intervals for free thyroxine and thyroid-stimulating hormone. Clin Chim Acta 2010;411:250–2.
	8	Ghoshal AK, Soldin SJ. Evaluation of the Dade Behring Dimension RxL: integrated chemistry system-pediatric reference ranges. Clin Chim Acta 2003;331:135–46.
	9	Kulasingam V, Jung BP, Blasutig IM, et al. Pediatric reference intervals for 28 chemistries and immunoassays on the Roche cobas® 6000 analyzer—A CALIPER pilot study. Clin Biochem 2010;43:1045–50.
Method(s)	1	Delphia Immunofluorescent TSH Kit (Pharmacia ENI Diagnostics Inc., Columbia, MD).
	2	Abbott IMx analyzer (Abbott Laboratories, Abbott Park, IL).
	3	Immuno I (Bayer Corp., Tarrytown, NY).
	4	ACS 180 using Chiron Diagnostics TSH-3kit (Chiron Diagnostics Corp., East Walpole, MA).
	5	Chemiluminescent immunoassay, Vitros ECi (Ortho-Clinical Diagnostics, Raritan, NJ).
	6	Abbott Architect ci8200 (Abbott Diagnostics, Abbott Park, IL).
	7	Abbott Architect ci8200 (Abbott Diagnostics, Abbott Park, IL).
	8	The TSH method is a one-step enzyme immunoassay based on the "sandwich" principle. Siemens Dimension RxL Analyzer (Siemens Healthcare Diagnostics, Deerfield, IL).
	9	Roche cobas® 6000 analyzer (Roche Diagnostics Limited, West Sussex, UK).

Comment(s)	1	Study used hospitalized patients and a computerized approach adapted from the Hoffmann technique. Values are 97.5th percentile. Lower limit of detection 0.1 IU/mL.
	2,3	Study used hospitalized patients and a computerized approach adapted from the Hoffmann technique. Values are 2.5–97.5th percentiles. Lower limit of detection 0.03 IU/mL.
	4	Study used hospitalized patients and a computerized approach adapted from the Hoffmann technique. Values are 2.5–97.5th percentiles. Lower limit of detection 0.003 IU/mL.
	5	*Ages ranged from 1 h to 8 y with a total of 119 specimens.
	6	1459 serum/plasma specimens from children attending select outpatient clinics were collected from the five age groups noted above. Values are 2.5–97.5th percentiles.
	7	Study encompassed 6023 children (3369 females and 2654 males). A poor correlation was observed between FT_4 and log TSH.
	8	**Numbers not provided. Reference ranges were obtained by comparing results from previously published data and using regression equations.
	9	Approximately 600 outpatient samples from a pediatric population deemed to be metabolically stable were subdivided into five age classes ranging from 0 to 20 years of age and further partitioned by gender. Values are 2.5–97.5th percentiles.

THYROXINE (T$_4$)

Test	Age	Male n	Male μg/dL	Male nmol/L	Female n	Female μg/dL	Female nmol/L
1	1–30 d	108	3.0–14.4	39–185	116	3.0–13.4	39–172
	1–12 mo	98	5.3–16.3	68–210	95	4.6–13.4	59–172
	1–5 y	151	5.5–11.4	71–147	197	6.3–12.8	81–165
	6–10 y	151	5.4–10.6	69–136	146	5.3–10.8	68–139
	11–15 y	171	4.5–10.3	58–133	160	4.9–10.0	63–129
	16–18 y	141	4.9–8.8	63–113	143	5.1–10.0	66–129
2	1–30 d	100	5.9–21.5	76–276	106	6.3–21.5	81–276
	31–364 d	107	6.4–13.9	82–179	96	4.9–13.7	63–176
	1–3 y	149	7.0–13.1	90–169	132	7.1–14.1	91–180
	4–6 y	139	6.1–12.6	79–162	117	7.2–14.0	93–180
	7–12 y	129	6.7–13.4	86–172	134	6.1–12.1	79–156
	13–15 y	145	4.8–11.5	62–148	108	5.8–11.2	75–144
	16–18 y	49	5.9–11.5	76–148	112	5.2–13.2	67–170
3	1–30 d	108	3.4–12.6	44–163	116	3.4–11.8	44–152
	1–12 mo	98	5.4–14.1	68–182	95	4.7–11.8	60–152
	1–5 y	151	5.3–10.2	69–131	197	6.0–11.3	78–146
	6–10 y	151	5.3–9.5	69–122	146	5.2–9.7	67–125
	11–15 y	171	4.6–9.2	59–119	160	4.9–9.0	63–116
	16–18 y	141	4.9–8.1	63–104	143	5.1–9.0	66–116
4	1–30 d	*	3.4–14.5	44–187	*	3.5–13.5	45–174
	1–12 mo		5.6–16.4	72–211		5.0–13.5	64–174
	1–5 y		5.9–11.6	76–149		6.7–12.9	86–166
	6–10 y		5.7–10.8	73–139		5.6–11.0	72–142
	11–15 y		4.9–10.5	63–135		5.3–10.2	68–131
	16–18 y		5.2–9.1	68–117		5.5–10.2	71–131

Test	Age	Male and Female		
		n	µg/dL	nmol/L
5	0–3 d	**	8.0–20.0	103–258
	3–30 d		5.0–15.0	64–193
	31–365 d		6.0–14.0	77–180
	1–5 y		4.5–11.0	58–142
	6–18 y		4.5–10.0	58–129
6	0–<1 y	67	5.5–14.4	71–186

	Male			Female		
Age	n	µg/dL	nmol/L	n	µg/dL	nmol/L
1–≤5 y	46	5.5–14.4	57–151	58	5.5–14.4	72–169

Male and Female			
Age	n	µg/dL	nmol/L
>5–≤20 y	284	4.4–12.1	57–157

Specimen Type(s)	1–4,6	Plasma/serum
	5	Serum
Reference(s)	1	Soldin SJ, Morales A, Albalos F, et al. Pediatric reference ranges on the Abbott IMx analyzer for FSH, LH, prolactin, TSH, T_4, T_3, free T_4, free T_4, Tuptake, IgE and ferritin. Clin Biochem 1995;28:603–6.
	2	Soldin SJ, Cook J, Beatey J, et al. Pediatric reference ranges for thyroxine and triiodothyronine uptake. Clin Chem 1992;38:960. (Abstract)
	3	Murthy JN, Hicks JM, Soldin SJ. Evaluation of the Technicon Immuno I Random Access Immunoassay Analyzer and calculation of pediatric reference ranges for endocrine tests, T-uptake, and ferritin. Clin Biochem 1995;28:181–5.
	4	Ghoshal AK, Soldin SJ. Evaluation of the Dade Behring Dimension RxL: integrated chemistry system-pediatric reference ranges. Clin Chim Acta 2003;331:135–46.
	5	Dugaw KA, Jack RM, Rutledge J. Pediatric reference ranges for TSH, free T_4, total T_4, total T_3 and T_3 uptake on the Vitros ECi analyzer. Chem Chem 2001;47:A108. (Abstract)
	6	Kulasingam V, Jung BP, Blasutig IM, et al. Pediatric reference intervals for 28 chemistries and immunoassays on the Roche cobas® 6000 analyzer—A CALIPER pilot study. Clin Biochem 2010;43:1045–50.
Method(s)	1	Abbott IMx Analyzer (Abbott Diagnostics, Inc., Abbott Park, IL).
	2	T_4: Gamma Coat ™[125] (Baxter-Travenol Diagnostics, Inc., Cambridge, MA).
	3	Immuno I (Bayer Corp., Tarrytown, NY).
	4	The thyroxine method is an adaptation of the EMIT® homogeneous enzyme immunoassay technology. Siemens Dimension RxL Analyzer (Siemens Healthcare Diagnostics, Deerfield, IL).
	5	Chemiluminescent immunoassay, Vitros ECi (Ortho-Clinical Diagnostics, Raritan, NJ).
	6	Roche cobas® 6000 analyzer (Roche Diagnostics Limited, West Sussex, UK).
Comment(s)	1–3	Study used hospitalized patients and a computerized approach adapted from the Hoffmann technique. Values are 2.5–97.5th percentiles.
	4	*Numbers not provided. Reference ranges were obtained by comparing results from previously published data and using regression equations.
	5	**Ages ranged from 1 h to 18 y with a total of 119 specimens.
	6	Approximately 600 outpatient samples from a pediatric population deemed to be metabolically stable were subdivided into five age classes ranging from 0 to 20 years of age and further partitioned by gender. Values are 2.5–97.5th percentiles.

THYROXINE BINDING GLOBULIN (TBG)

Test	Age	Male n	Male mg/L	Female n	Female mg/L
1	1–12 mo	189	16.2–32.9	138	17.7–32.0
	1–3 y	167	16.4–32.0	112	19.3–33.8
	4–6 y	100	16.6–29.8	91	18.3–30.8
	7–12 y	184	16.5–28.8	133	15.0–29.2
	13–18 y	158	13.4–25.6	130	13.7–28.7
2	Cord blood	*	19–39	*	19–39
	1–11 mo		16–36		17–37
	1–9 y		12–28		15–27
	10–19 y		14–26		14–30

Specimen Type(s)	1	Plasma
	2	Plasma/serum
Reference(s)	1	Hicks JM, Godwin ID, Beatey J, et al. Pediatric reference ranges for thyroid binding globulin. Clin Chem 1993:39;1172. (Abstract)
	2	Levy RP, Marshall JS, Velayo NL. Radioimmunoassay of human thyroxine binding globulin (TBG). J Clin Endocrinol Metab 1971;32:372–81.
Method(s)	1	Corning TBG[125]I–Radioimmunoassay. (Ciba Corning Diagnostics Corp., East Walpole, MA).
	2	Direct RIA using goat antihuman TBG and ^{125}I–labeled human TBG. A second antibody is used for the bound/free separation phase.
Comment(s)	1	Study used hospitalized patients and a computerized approach to removing outliers. Values are 2.5–97.5th percentiles.
	2	Study used healthy children.

*For numbers at each age group, see Meites S, ed. Pediatric clinical chemistry, 3rd ed. Washington, DC: AACC Press, 1989:254.

THYROXINE, FREE (FREE T$_4$)

Test	Age	Male			Female		
		n	ng/dL	pmol/L	n	ng/dL	pmol/L
1	1–3 d	24	0.80–2.78	10–36	38	0.88–1.93	11–25
	4–30 d	73	0.48–2.32	6–30	62	0.61–1.93	8–25
	1–12 mo	52	0.76–2.00	10–26	54	0.88–1.84	11–24
	1–5 y	100	0.90–1.59	12–21	117	1.02–1.72	13–22
	6–10 y	104	0.81–1.68	10–22	101	0.82–1.58	11–20
	11–15 y	101	0.92–1.57	12–20	100	0.79–1.49	10–19
	16–18 y	110	0.92–1.53	12–20	101	0.83–1.44	11–19
2	1–3 d	24	1.16–2.95	15–38	38	1.09–2.09	14–27
	4–30 d	73	0.78–2.25	10–29	62	0.85–2.09	11–27
	1–12 mo	52	1.00–2.17	13–28	54	1.09–2.02	14–26
	1–5 y	100	1.16–1.71	15–22	117	1.24–1.86	16–24
	6–10 y	104	1.09–1.86	14–24	101	1.09–1.71	14–22
	11–15 y	101	1.16–1.71	15–22	100	1.00–1.63	13–21
	16–18 y	110	1.16–1.71	15–22	101	1.09–1.63	14–21
3	<2 mo	189	0.78–1.83	10.1–23.6	160	0.69–1.85	8.9–23.9
	2–<12 mo	125	0.71–1.56	9.2–20.1	138	0.69–1.49	8.9–19.2
	12–24 mo	190	0.82–1.32	10.6–17.0	202	0.73–1.41	9.4–18.2
	2–<5 y	286	0.80–1.32	10.3–17.0	253	0.79–1.38	10.2–17.8
	5–<10 y	516	0.78–1.29	10.1–16.6	655	0.77–1.32	9.9–17.0
	10–<15	722	0.69–1.23	8.9–15.9	1003	0.66–1.22	8.5–15.7
	15–<20 y	507	0.67–1.22	8.6–15.7	802	0.67–1.22	8.6–15.7
4	0–12 mo	36	0.92–1.83	11.9–23.6	43	0.85–1.59	11.0–20.6
	1–5 y	101	0.86–1.62	11.0–20.8	93	0.91–1.44	11.7–18.6

	Age	Male and Female					
		n	ng/dL			pmol/L	
	6–10 y	139	0.84–1.47			10.9–19.0	
	11–14 y	161	0.78–1.31			10.0–16.9	
	15–20 y	163	0.79–1.34			10.2–17.3	

		Ultrafiltration at 25°C					
5	1 mo–18 y	1000	0.8–2.1			10–27	

Test	Age	n	Male and Female ng/dL	Male and Female pmol/L
6	0–3 d	*	2.0–5.0	25.7–64.3
	3–30 d		0.9–2.2	11.6–28.3
	31 d–18 y		0.8–2.0	10.3–25.7

Test	Age	n	Male ng/dL	Male pmol/L	Female ng/dL	Female pmol/L
7	1–3 d	**	0.97–1.87	12.5–24.1	0.93–1.44	12.0–18.5
	4–30 d		0.78–1.52	10.0–19.6	0.81–1.44	10.4–18.5
	1–12 mo		0.89–1.48	11.5–19.1	0.93–1.40	12.0–18.0
	1–5 y		0.97–1.25	12.5–16.1	1.01–1.32	13.0–17.0
	6–10 y		0.93–1.32	12.0–17.0	0.93–1.25	12.0–16.1
	11–15 y		0.97–1.25	12.5–16.1	0.89–1.20	11.5–15.4
	16–18 y		0.97–1.25	12.5–16.1	0.93–1.25	12.0–16.1

Test	Age	n	Male and Female — Ultrafiltration at 37 °C ng/dL	Male and Female — Ultrafiltration at 37 °C pmol/L
8	1 mo–<1 y	140	1.3–2.8	16.8–36.1
	1–<3 y	145	1.3–2.4	16.8–31.0
	3–<8 y	129	1.3–2.4	16.8–31.0

Male Age	n	Male Ultrafiltration at 37 °C ng/dL	Male Ultrafiltration at 37 °C pmol/L	Female Age	n	Female Ultrafiltration at 37 °C ng/dL	Female Ultrafiltration at 37 °C pmol/L
3–<8 y	119	1.3–2.4	16.8–31.0	8–<11 y	109	1.3–2.4	16.8–31.0
8–<12 y	152	1.3–2.4	16.8–31.0	11–<14 y	137	1.3–2.4	16.8–31.0
12–<15 y	135	1.3–2.4	16.8–31.0	14–<16 y	119	1.3–2.4	16.8–31.0
15–<18 y	121	1.3–2.4	16.8–31.0	16–<18 y	120	1.3–2.4	16.8–31.0

Specimen Type(s)	1–4,7	Plasma/serum
	6,8	Serum
Reference(s)	1	Soldin SJ, Morales A, Albalos F, et al. Pediatric reference ranges on the Abbott IMx for FSH, LH, prolactin, TSH, T_4, T_3, free T_4, free T_4, T-uptake, IgE and ferritin. Clin Biochem 1995;28:603–6.
	2	Murthy JN, Hicks, JM, Soldin SJ. Evaluation of the Technicon Immuno I Random Access Immunoassay Analyzer and calculation of pediatric reference ranges for endocrine tests, T-uptake, and ferritin. Clin Biochem 1995;28:181–5.
	3	Soldin SJ, Cheng LL, Lam LY, et al. Comparison of FT4 with log TSH on the Abbott Architect ci8200: pediatric reference intervals for free thyroxine and thyroid-stimulating hormone. Clin Chim Acta 2010;411:250–2.
	4	Chan MK, Seiden-Long I, Aytekin M, et al. Canadian Laboratory Initiative on Reference Interval Database (CALIPER): pediatric reference intervals for an integrated clinical chemistry and immunoassay analyzer, Abbott ARCHITECT ci8200. Clin Biochem 2009;42:885–91.
	5	Gu J, Soldin SJ. Simultaneous quantification of free triiodothyronine and free thyroxine by tandem mass spectrometry. Clin Chem 2007;53:A190. (Abstract)
	6	Dugaw KA, Jack RM, Rutledge J. Pediatric reference ranges for TSH, free T4, total T4, total T3 and T3 uptake on the Vitros ECi analyzer. Clin Chem 2001;47:A108. (Abstract)
	7	Ghoshal AK, Soldin SJ. Evaluation of the Dade Behring Dimension RxL: integrated chemistry system-pediatric reference ranges. Clin Chim Acta 2003;331:135–46.
	8	Soldin OP, Jang M, Guo T, Soldin SJ. Pediatric reference intervals for free thyroxine and free triiodothyronine. Thyroid 2009;19:699–702.
Method(s)	1	Abbott IMx Analyzer (Abbott Diagnostics, Abbott Park, IL).
	2	Immuno I (Bayer Corp., Tarrytown, NY).
	3	Abbott Architect ci8200 (Abbott Diagnostics, Abbott Park, IL).
	4	Abbott Architect ci8200 (Abbott Diagnostics, Abbott Park, IL).
	5	Isotope dilution tandem mass spectrometry (API-5000, Sciex, Concord, Canada) in the negative mode. Ultrafiltration was performed at 25 °C. Multiply by 1.5 to compare to ultrafiltration performed at 37 °C.
	6	Chemiluminescent immunoassay, Vitros ECi (Ortho-Clinical Diagnostics, Raritan, NJ).
	7	The FT_4 method for the Dimension® RxL system with the heterogeneous immunoassay module is a two-step enzyme immunoassay based on a sequential competitive format. Siemens Dimension RxL Analyzer (Siemens Healthcare Diagnostics, Deerfield, IL).
	8	Isotope dilution liquid chromatography tandem mass spectrometry (LC/MS/MS) with deuterium-labeled internal standards. Ultrafiltration was performed at 37 °C.

Comment(s)	1,2	Study used hospitalized patients and a computerized approach adapted from the Hoffmann technique. Values are 2.5–97.5th percentiles.
	3	Study encompassed 6023 children (3369 females and 2654 males). FT_4 by this method correlates poorly with log TSH.
	4	1459 serum/plasma specimens from children attending select outpatient clinics were collected from the five age groups noted above. Values are 2.5–97.5th percentiles.
	5	Reference intervals obtained using the Hoffmann technique.
	6	*Ages ranged from 1 h to 18 y with a total of 119 specimens.
	7	**Numbers not provided. Reference intervals were obtained by comparing results from previously published data and using regression equations. FT_4 by this method correlates poorly with log TSH.
	8	Reference intervals were calculated for serum obtained from healthy children between January 1–June 30, 2008, from Children's National Medical Center and Georgetown University Medical Center Bioanalytical Core Laboratory. This is the first study to provide pediatric reference intervals of free thyroxine for children from birth to 18 y using LC/MS/MS. Values are 2.5–97.5th percentiles. FT_4 by this method correlates well with log TSH.

TRANSFERRIN

Test	Age	n	mg/dL	g/L
			Male and Female	
1	0–5 d	73	124–388	1.24–3.88
	1–3 y	51	190–302	1.90–3.02
	4–6 y	39	181–329	1.81–3.29
	7–9 y	39	196–314	1.96–3.14
	10–13 y	110	195–385	1.95–3.85
	14–19 y	78	203–386	2.03–3.86
2	0–12 mo	92	133–332	1.33–3.32
	1–5 y	146	204–366	2.04–3.66

		Male			Female		
	Age	n	mg/dL	g/L	n	mg/dL	g/L
	6–10 y	53	217–321	2.17–3.21	72	177–371	1.77–3.71
	11–14 y	67	181–353	1.81–3.53	92	200–367	2.00–3.67
	15–20 y	37	183–363	1.83–3.63	110	193–421	1.93–4.21
3	1–30 d	41	84–178	0.84–1.78	36	80–181	0.80–1.81
	31–182 d	88	92–283	0.92–2.83	70	111–269	1.11–2.69
	183–365 d	40	155–311	1.55–3.11	29	127–317	1.27–3.17
	1–3 y	157	171–318	1.71–3.18	142	130–332	1.30–3.32
	4–6 y	108	176–305	1.76–3.05	94	151–347	1.51–3.47
	7–9 y	91	130–307	1.30–3.07	91	162–320	1.62–3.20
	10–12 y	85	151–331	1.51–3.31	105	161–328	1.61–3.28
	13–15 y	82	149–325	1.49–3.25	84	168–340	1.68–3.40
	16–18 y	60	169–303	1.69–3.03	82	157–362	1.57–3.62

Test	Age	n	Male		Female	
			mg/dL	g/L	mg/dL	g/L
4	1–30 d	*	86–174	0.86–1.74	83–176	0.83–1.76
	31–182 d		94–271	0.94–2.71	111–258	1.11–2.58
	183–365 d		152–296	1.52–2.96	126–303	1.26–3.03
	1–3 y		167–304	1.67–3.04	129–317	1.29–3.17
	4–6 y		172–291	1.72–2.91	149–331	1.49–3.31
	7–9 y		129–293	1.29–2.93	159–305	1.59–3.05
	10–12 y		115–316	1.15–3.16	158–313	1.58–3.13
	13–15 y		147–310	1.47–3.10	164–324	1.64–3.24
	16–18 y		165–289	1.65–2.89	154–344	1.54–3.44

Specimen Type(s)	1–3	Plasma/serum
	4	Serum
Reference(s)		The above values have been adjusted from those published in references 1 and 3 to convert results to current IFCC units using IFCC guidelines/standards.
	1	Lockitch G, Halstead AC, Quigley G, et al. Age- and sex-specific pediatric reference intervals: study, design and methods illustrated by measurement of serum proteins with the Behring LN nephelometer. Clin Chem 1988;34:1618–21.
	2	Chan MK, Seiden-Long I, Aytekin M, et al. Canadian Laboratory Initiative on Reference Interval Database (CALIPER): pediatric reference intervals for an integrated clinical chemistry and immunoassay analyzer, Abbott ARCHITECT ci8200. Clin Biochem 2009;42:885–91.
	3	Soldin SJ, Hicks JM, Bailey J, et al. Pediatric reference ranges for total iron binding capacity and transferrin. Clin Chem 1997;43:S200. (Abstract)
	4	Ghoshal AK, Soldin SJ. Evaluation of the Dade Behring Dimension RxL: integrated chemistry system-pediatric reference ranges. Clin Chim Acta 2003;331:135–46.
Method(s)	1	Nephelometric, Behring LN nephelometer (Behring Diagnostics, Hoechst, Inc., Montreal, Canada).
	2	Abbott Architect ci8200 (Abbott Diagnostics, Abbott Park, IL).
	3	Nephelometric, Behring LN nephelometer (Behring Diagnostics, Inc., Westwood, MA).
	4	The method is a quantitative turbidimetric assay using endpoint detection based on the precipitation of transferrin by its polyclonal antibody. Siemens Dimension RxL Analyzer (Siemens Healthcare Diagnostics, Deerfield, IL).
Comment(s)	1	Normal healthy children. Values provided are 2.5–97.5th percentiles.
	2	1459 serum/plasma specimens from children attending select outpatient clinics were collected from the five age groups noted above. Values are 2.5–97.5th percentiles.
	3	Study used hospitalized patients and a computerized adaptation of the Hoffmann technique. Results are 2.5–97.5th percentiles.
	4	*Numbers not provided. Reference ranges were obtained by comparing results from previously published data and using regression equations.

TRANSFERRIN SATURATION

Age	Male n		Female n	
1–5 y*	44	0.07–0.44	44	0.07–0.44
6–9 y*	50	0.17–0.42	50	0.17–0.42
10–14 y	31	0.11–0.36	40	0.02–0.40
14–19 y	65	0.06–0.33	110	0.06–0.33

Specimen Type(s)	Plasma
Reference(s)	Lockitch G, Halstead AC, Wadsworth L, et al. Age- and sex-specific pediatric reference intervals and correlations for zinc, copper, selenium, iron, vitamins A and E, and related proteins. Clin Chem 1988;34:1625–8.
Method(s)	Transferrin was measured by nephelometry. Behring LN with Behring kit (Behringwerke, Marburg, Germany).
Comment(s)	The study population was healthy children. Non-parametric methods were used to determine the 0.025 and 0.975 fractiles. *Males and females were not studied separately. Transferrin saturation = iron (μmol/L) $\div$ TIBC.

TRIGLYCERIDES

Test	Age	Male			Female		
		n	mmol/L	mg/dL	n	mmol/L	mg/dL
1	0–7 d	149	0.24–2.06	21–182	142	0.32–1.88	28–166
	8–30 d	283	0.34–2.08	30–184	172	0.34–1.86	30–165
	31–90 d	247	0.45–1.98	40–175	171	0.40–3.19	35–282
	91–180 d	132	0.51–3.29	45–291	126	0.57–4.01	50–355
	181–365 d	286	0.51–5.66	45–501	261	0.41–4.87	36–431
2	1–3 y	49*	0.31–1.41	27–125	49*	0.31–1.41	27–125
	4–6 y	38*	0.36–1.31	32–116	38*	0.36–1.31	32–116
	7–9 y	72*	0.32–1.46	28–129	72*	0.32–1.46	28–129
	10–11 y	28	0.27–1.55	24–137	34	0.44–1.58	39–140
	12–13 y	32	0.27–1.64	24–145	40	0.42–1.47	37–130
	14–15 y	39	0.38–1.86	34–165	50	0.43–1.52	38–135
	16–19 y	41	0.38–1.58	34–140	68	0.42–1.58	37–140
3	Cord blood serum arterial	397**	0.10–1.04	9–92	397**	0.10–1.04	9–92
	Cord blood serum venous	397**	0.13–0.97	12–86	397**	0.13–0.97	12–86
4	0–7 d	***	0.21–1.97	19–174	***	0.29–1.80	26–159
	8–30 d		0.42–3.15	37–279		0.37–3.05	33–270
	31–90 d		0.47–3.15	42–279		0.38–3.84	34–340
	1–3 y		0.28–1.34	25–119		0.28–1.34	25–119
	4–6 y		0.34–1.24	30–110		0.34–1.24	30–110
	7–9 y		0.29–1.39	26–123		0.29–1.39	26–123
	10–11 y		0.25–1.48	22–131		0.42–1.51	37–134
	12–13 y		0.25–1.56	22–138		0.40–1.40	35–124
	14–15 y		0.36–1.78	32–158		0.41–1.46	36–129
	16–19 y		0.36–1.51	32–134		0.40–1.51	35–134

		Male and Female		
Test	Age	n	mmol/L	mg/dL
5	0–12 mo	128	0.55–3.84	42–295
	1–5 y	152	0.45–2.73	40–242
	6–10 y	134	0.46–3.19	41–282
	11–14 y	154	0.46–3.53	41–312
	15–20 y	181	0.47–2.42	42–214
6	0–<1 y	94	0.59–2.63	52–233
	≥1–≤20 y	456	0.43–2.75	38–243

Specimen Type(s)	1	Serum
	2–6	Plasma/serum
Reference(s)	1	Soldin SJ, Morse AS. Pediatric reference ranges for calcium and triglycerides in children <1 year old using the Vitros 500 Analyzer. Clin Chem 1998;44:A16. (Abstract)
	2	Lockitch G, Halstead AC, Albersheim S, et al. Age- and sex-specific pediatric reference intervals for biochemistry analytes as measured with the Ektachem-700 analyzer. Clin Chem 1988;34:1622–5.
	3	Perkins SL, Livesey JF, Belcher J. Reference intervals for 21 clinical chemistry analytes in arterial and venous umbilical cord blood. Clin Chem 1993;39:1041–4.
	4	Ghoshal AK, Soldin SJ. Evaluation of the Dade Behring Dimension RxL: integrated chemistry system-pediatric reference ranges. Clin Chim Acta 2003;331:135–46.
	5	Chan MK, Seiden-Long I, Aytekin M, et al. Canadian Laboratory Initiative on Reference Interval Database (CALIPER): pediatric reference intervals for an integrated clinical chemistry and immunoassay analyzer, Abbott ARCHITECT ci8200. Clin Biochem 2009;42:885–91.
	6	Kulasingam V, Jung BP, Blasutig IM, et al. Pediatric reference intervals for 28 chemistries and immunoassays on the Roche cobas® 6000 analyzer—A CALIPER pilot study. Clin Biochem 2010;43:1045–50.
Method(s)	1,2	Glycerol phosphate oxidase on Vitros 700 analyzer (Ortho-Clinical Diagnostics, Raritan, NJ).
	3	Hitachi 737 with Boehringer Mannheim reagents (Boehringer Mannheim Canada, Montreal, Canada).
	4	The sample is pre-incubated with lipase enzyme reagent, which converts triglycerides into free glycerol and fatty acids. The liberated glycerol is determined enzymatically using glycerol dehydrogenase and NAD. Siemens Dimension RxL Analyzer (Siemens Healthcare Diagnostics, Deerfield, IL).
	5	Abbott Architect ci8200 (Abbott Diagnostics, Abbott Park, IL).
	6	Roche cobas® 6000 analyzer (Roche Diagnostics Limited, West Sussex, UK).

Comment(s)	1	The study used plasma/serum obtained from hospitalized patients and employed Chauvenet's criteria to remove outliers and a computerized approach adapted from the Hoffmann technique to obtain the 2.5–97.5th percentiles. Fasting samples were not obtained in these neonates and infants, which accounts for the somewhat elevated 97.5th percentiles.
	2	A healthy population of children 1–19 y was studied. Results are 2.5–97.5th percentiles. *No significant differences were found for males and females. These ranges were therefore derived from combined data.
	3	Results are 2.5–97.5th percentiles. **No significant differences were found for males and females. These ranges were therefore derived from combined data.
	4	***Numbers not provided. Reference ranges were obtained by comparing results from previously published data and using regression equations.
	5	1459 serum/plasma specimens from children attending select outpatient clinics were collected from the five age groups noted above. Values are 2.5–97.5th percentiles.
	6	Approximately 600 outpatient samples from a pediatric population deemed to be metabolically stable were subdivided into five age classes ranging from 0 to 20 years of age and further partitioned by gender. Values are 2.5–97.5th percentiles.

TRIIODOTHYRONINE (T$_3$)

Test	Age	Male			Female		
		n	ng/dL	nmol/L	n	ng/dL	nmol/L
1	1–30 d	50	15–210	0.2–3.2	47	15–200	0.2–3.1
	1–12 mo	111	95–275	1.5–4.2	100	50–264	0.8–4.1
	1–5 y	101	80–253	1.2–3.9	115	126–258	1.9–4.0
	6–10 y	99	96–232	1.5–3.6	99	104–227	1.6–3.5
	11–15 y	97	73–199	1.1–3.1	100	96–211	1.5–3.2
	16–18 y	100	69–201	1.1–3.1	104	91–164	1.4–2.5
2	1–30 d	227	29–179	0.4–2.7	173	34–173	0.5–2.7
	1–3 y	115	58–190	0.9–2.9	127	54–199	0.8–3.1
	4–6 y	112	59–177	0.9–2.7	105	58–173	0.9–2.7
	7–12 y	114	52–192	0.8–2.9	110	53–171	0.8–2.6
	13–18 y	113	64–157	1.0–2.4	121	40–156	0.6–2.4
3	1–30 d	50	51–184	0.7–2.9	47	48–177	0.7–2.8
	1–12 mo	111	103–229	1.6–3.5	100	73–221	1.1–3.4
	1–5 y	101	93–213	1.4–3.3	115	126–216	1.9–3.4
	6–10 y	99	104–198	1.6–3.0	99	110–195	1.7–3.0
	11–15 y	97	88–176	1.3–2.7	100	104–184	1.6–2.9
	16–18 y	100	86–178	1.3–2.8	104	101–151	1.5–2.4
4	0–<12 mo	74	48–320	0.7–4.9	60	36–320	0.5–4.9
	1–<5 y	130	90–285	1.4–4.4	80	54–304	0.8–4.6
	5–<9 y	123	60–290	0.9–4.5	96	64–272	1.0–4.2
	9–<13 y	178	65–270	1.0–4.2	159	62–248	0.9–3.8
	13–<16 y	178	60–228	0.9–3.5	289	52–198	0.8–3.0
	16–<19 y	114	38–208	0.6–3.2	206	28–202	0.4–3.1
5	0–12 mo	38	61–204	1.0–3.1	43	102–229	1.6–3.5

	Male and Female		
Age	n	ng/dL	nmol/L
1–5 y	194	106–203	1.6–3.1
6–10 y	141	104–183	1.6–2.8
11–14 y	162	68–186	1.0–2.9
15–20 y	164	71–175	1.1–2.7

Test	Age	Male and Female			
		n	ng/dL	nmol/L	
6	0–3 d	*	60–300	0.9–4.7	
	4–365 d		90–260	1.4–4.0	
	1–6 y		90–240	1.4–3.7	
	7–11 y		90–230	1.4–3.6	
	12–18 y		100–210	1.5–3.3	
7	0–≤10 y	297	82–282	1.3–4.3	

Test	Age	Male			Female		
		n	ng/dL	nmol/L	n	ng/dL	nmol/L
	>10–≤15 y	61	80–233	1.2–3.6	58	60–209	0.9–3.2
	>15–≤20 y	27	71–212	1.1–3.3	35	61–151	0.9–2.3

Specimen Type(s)	1–5,7	Plasma/serum
	6	Serum
Reference(s)	1	Soldin SJ, Morales A, Albalos F, et al. Pediatric reference ranges on the Abbott IMx for FSH, LH, prolactin, TSH, T_4, T_3, free T_4, free T_3, T-uptake, IgE, and ferritin. Clin Biochem 1995;28:603–6.
	2	Soldin SJ, Hicks JM, Bailey J, et al. Pediatric reference ranges for triiodothyronine. Clin Chem 1997;43:S199. (Abstract)
	3	Murthy JN, Hicks JM, Soldin SJ. Evaluation of the Technicon Immuno I Random Access Immunoassay Analyzer and calculation of pediatric reference ranges for endocrine tests, T-uptake, and ferritin. Clin Biochem 1995;28:181–5.
	4	Soldin OP, Hoffman EG, Waring MA, Soldin SJ. Pediatric reference intervals for FSH, LH, estradiol, T_3, free T_3, cortisol, and growth hormone on the DPC IMMULITE 1000. Clin Chim Acta 2005;355:205–10.
	5	Chan MK, Seiden-Long I, Aytekin M, et al. Canadian Laboratory Initiative on Reference Interval Database (CALIPER): pediatric reference intervals for an integrated clinical chemistry and immunoassay analyzer, Abbott ARCHITECT ci8200. Clin Biochem 2009;42:885–91.
	6	Dugaw KA, Jack RM, Rutledge J. Pediatric reference ranges for TSH, free T_4, total T_4, total T_3 and T_3 uptake on the Vitros ECi analyzer. Chem Chem 2001;47:A108. (Abstract)
	7	Kulasingam V, Jung BP, Blasutig IM, et al. Pediatric reference intervals for 28 chemistries and immunoassays on the Roche cobas® 6000 analyzer—A CALIPER pilot study. Clin Biochem 2010;43:1045–50.
Method(s)	1	Abbott IMx Analyzer (Abbott Diagnostics, Inc., Abbott Park, IL).
	2	T_3 (Diagnostic Products, Inc. Los Angeles, CA).
	3	Immuno I (Bayer Corp., Tarrytown, NY).
	4	Siemens IMMULITE® 1000 analyzer (Siemens Healthcare Diagnostics, Deerfield, IL).
	5	Abbott Architect ci8200 (Abbott Diagnostics, Abbott Park, IL).
	6	Chemiluminescent immunoassay, Vitros ECi (Ortho-Clinical Diagnostics, Raritan, NJ).
	7	Roche cobas® 6000 analyzer (Roche Diagnostics Limited, West Sussex, UK).
Comment(s)	1–4	Study used hospitalized patients and a computerized approach adapted from the Hoffmann technique. Values are 2.5–97.5th percentiles.
	5	1459 serum/plasma specimens from children attending select outpatient clinics were collected from the five age groups noted above. Values are 2.5–97.5th percentiles.
	6	*Ages ranged from 1 h to 18 y with a total of 119 specimens.
	7	Approximately 600 outpatient samples from a pediatric population deemed to be metabolically stable were subdivided into five age classes ranging from 0 to 20 years of age and further partitioned by gender. Values are 2.5–97.5th percentiles.

TRIIODOTHYRONINE, FREE (FREE T$_3$)

Test	Age	Male n	Male ng/dL	Male pmol/L	Female n	Female ng/dL	Female pmol/L
1	1–3 d	24	0.14–0.48	2.2–7.4	26	0.14–0.54	2.2–8.3
	4–30 d	73	0.14–0.55	2.2–8.4	62	0.15–0.50	2.3–7.7
	1–12 mo	52	0.20–0.69	3.1–10.6	52	0.25–0.65	3.8–10.0
	1–5 y	100	0.24–0.67	3.7–10.3	99	0.30–0.60	4.6–9.2
	6–10 y	104	0.29–0.60	4.4–9.2	101	0.27–0.62	4.1–9.5
	11–15 y	102	0.31–0.59	4.8–9.1	102	0.26–0.57	4.0–8.8
	16–18 y	101	0.35–0.57	5.4–8.8	98	0.28–0.52	4.3–8.0
2	0–6 y	62	0.40–0.71	6.1–10.9	62	0.40–0.71	6.1–10.9
	7–12 y	43	0.33–0.80	5.1–12.3	43	0.33–0.80	5.1–12.3
	13–18 y	33	0.23–0.75	3.5–11.5	33	0.23–0.75	3.5–11.5
3	11–14 y	46	0.24–0.45	3.7–6.9	116	0.22–0.42	3.4–6.5
	15–20 y	37	0.19–0.44	2.9–6.8	130	0.22–0.39	3.3–5.8

	Male and Female		
Age	n	ng/dL	pmol/L
0–12 mo	81	0.22–0.49	3.4–7.6
1–5 y	193	0.28–0.47	4.3–7.2
6–10 y	141	0.28–0.44	4.4–6.8

Test	Age	n	ng/dL	pmol/L
4	0–<19 y	90	0.11–0.34	1.7–5.2
			Ultrafiltration at 25 °C	
5	1 mo–18 y	1000	0.09–0.40	1.4–6.1
			Ultrafiltration at 37 °C	
6	1 mo–<8 y	124	0.15–0.60	2.3–9.2

Male				Female			
Ultrafiltration at 37 °C				Ultrafiltration at 37 °C			
Age	n	ng/dL	pmol/L	Age	n	ng/dL	pmol/L
8–<12 y	144	0.15–0.67	2.3–10.3	8–<13 y	122	0.15–0.67	2.3–10.3
12–<14 y	140	0.15–0.62	2.3–9.5	13–<15 y	145	0.15–0.63	2.3–9.7
14–<16 y	102	0.15–0.60	2.3–9.2	15–<17 y	187	0.15–0.60	2.3–9.2
16–<18 y	86	0.15–0.67	2.3–10.3	17–<18 y	57	0.15–0.61	2.3–9.4

Specimen Type(s)	1–3,5	Plasma/serum
	4,6	Serum
Reference(s)	1	Soldin SJ, Morales A, Albalos F, et al. Pediatric reference ranges on the Abbott IMx for FSH, LH, prolactin, TSH, T_4, T_3, free T_4, free T_3, T-uptake, IgE, and ferritin. Clin Biochem 1995;28:603–6.
	2	Butler J, Moore P, Mieli-Vergani G, et al. Serum free thyroxine and free triiodothyronine in normal children. Ann Clin Biochem 1988;25:536–9.
	3	Chan MK, Seiden-Long I, Aytekin M, et al. Canadian Laboratory Initiative on Reference Interval Database (CALIPER): pediatric reference intervals for an integrated clinical chemistry and immunoassay analyzer, Abbott ARCHITECT ci8200. Clin Biochem 2009;42:885–91.
	4	Soldin OP, Hoffman EG, Waring MA, Soldin SJ. Pediatric reference intervals for FSH, LH, estradiol, T_3, free T_3, cortisol, and growth hormone on the DPC IMMULITE 1000. Clin Chim Acta 2005;355:205–10.
	5	Gu J, Soldin SJ. Simultaneous quantification of free triiodothyronine and free thyroxine by tandem mass spectrometry. Clin Chem 2007;53:A190. (Abstract)
	6	Soldin OP, Jang M, Guo T, Soldin SJ. Pediatric reference intervals for free thyroxine and free triiodothyronine. Thyroid 2009;19:699–702.
Method(s)	1	Abbott IMx Analyzer (Abbott Diagnostics, Inc., Abbott Park, IL).
	2	RIA Amerlex (Amersham International plc, Cardiff, UK).
	3	Abbott Architect ci8200 (Abbott Diagnostics, Abbott Park, IL).
	4	Siemens IMMULITE® 1000 analyzer (Siemens Healthcare Diagnostics, Deerfield, IL).
	5	Isotope dilution tandem mass spectrometry (API-5000, Sciex, Concord, Canada) in the negative mode. Results reflect ultrafiltration performed at 25 °C.
	6	Isotope dilution liquid chromatography tandem mass spectrometry (LC/MS/MS) with deuterium-labeled internal standards. Results reflect ultrafiltration performed at 37 °C.
Comment(s)	1,4	Study used hospitalized patients and a computerized approach adapted from the Hoffmann technique. Values are 2.5–97.5th percentiles.
	2	Normal school children were used in this study. Data were log-transformed and the mean and standard deviation were calculated.
	3	1459 serum/plasma specimens from children attending select outpatient clinics were collected from the five age groups noted above. Values are 2.5–97.5th percentiles.
	5	Reference intervals obtained using the Hoffmann technique.
	6	Reference intervals were calculated for serum obtained from healthy children between January 1–June 30, 2008, from Children's National Medical Center and Georgetown University Medical Center Bioanalytical Core Laboratory. This is the first study to provide pediatric reference intervals of free thyroxine for children from birth to 18 y using LC/MS/MS. Values are 2.5–97.5th percentiles.

TRIIODOTHYRONINE UPTAKE TEST (T₃U)

Test	Age	Male		Female	
		n	%	n	%
1	1–365 d	96	23–34	95	23–36
	1–3 y	147	24–35	127	24–36
	4–6 y	124	24–34	116	24–35
	7–12 y	127	24–33	136	22–35
	13–15 y	142	25–37	111	23–37
	16–18 y	47	24–38	114	23–35
		Male		Female	
Test	Age	n	T-Uptake Units	n	T-Uptake Units
2	1–30 d	34	0.26–1.66	44	0.41–1.44
	1 mo–5 y	99	0.72–1.35	141	0.81–1.36
	6–10 y	96	0.74–1.22	98	0.78–1.18
	11–19 y	113	0.58–1.23	108	0.76–1.25
		Male		Female	
Test	Age	n	Ratio	n	Ratio
3	0–1 mo	34	0.71–1.25	44	0.88–1.08
	1 mo–5 y	99	0.89–1.13	141	0.92–1.13
	6–10 y	96	0.84–1.08	98	0.82–1.10
	11–18 y	113	0.88–1.08	108	0.77–1.16
		Male and Female			
Test	Age	n	%		
4	0–30 d	*	24–48		
	31 d–18 y		26–39		

Specimen Type(s)	1–3	Plasma/serum
	4	Serum
Reference(s)	1	Soldin SJ, Cook J, Beatey J, et al. Pediatric reference ranges for thyroxine and triiodothyronine. Clin Chem 1992;38:960. (Abstract)
	2	Soldin SJ, Morales A, Albalos F, et al. Pediatric reference ranges on the Abbott IMx for FSH, LH, prolactin, TSH, T_4, T_3, free T_4, free T_3, T-uptake, IgE, and ferritin. Clin Biochem 1995;28:603–6.
	3	Murthy JN, Hicks JM, Soldin SJ. Evaluation of the Technicon Immuno I Random Access Immunoassay Analyzer and calculation of pediatric reference ranges for endocrine tests, T-uptake, and ferritin. Clin Biochem 1995;28:181–5.
	4	Dugaw KA, Jack RM, Rutledge J. Pediatric reference ranges for TSH, free T_4, total T_4, total T_3 and T_3 uptake on the Vitros ECi analyzer. Chem Chem 2001;47:A108. (Abstract)
Method(s)	1	RIA Quantimmune® 11 (Bio-Rad, Bio-Rad Laboratories, Richmond, CA).
	2	Abbott IMx (Abbott Laboratories, Abbott Park, IL).
	3	Immuno I (Bayer Corp., Tarrytown, NY).
	4	Chemiluminescent immunoassay, Vitros ECi (Ortho-Clinical Diagnostics, Raritan, NJ).
Comment(s)	1–3	Study used hospitalized patients and a computerized approach adapted from the Hoffmann technique to obtain the 2.5–97.5th percentiles.
	4	*Ages ranged from 1 h to 18 y with a total of 119 specimens.

TROPONIN I

Test	Age	n	97.5th percentile lg/L
	Male and Female		
1	0–30 d	97	<4.8
	31–90 d	46	<0.4
	3–6 mo	91	<0.3
	7–12 mo	53	<0.2
	1–18 y	57	<0.1
2	0–30 d	*	<8.4
	31–90 d		<0.7
	3–6 mo		<0.5
	7–12 mo		<0.3
	1–18 y		<0.1

Specimen Type(s)	1,2	Plasma/serum
Reference(s)	1	Soldin SJ, Murthy JN, Agarwalla PK, et al. Pediatric reference ranges for creatine kinase, CKMB, troponin I, and cortisol. Clin Biochem 1999;32:77–80.
	2	Ghoshal AK, Soldin SJ. Evaluation of the Dade Behring Dimension RxL: integrated chemistry system-pediatric reference ranges. Clin Chim Acta 2003;331:135–46.
Method(s)	1	Bayer Immuno I with Bayer reagents (Bayer Corp., Tarrytown, NY).
	2	The TROP method is a one-step enzyme immunoassay based on the "sandwich" principle. Siemens Dimension RxL Analyzer (Siemens Healthcare Diagnostics, Deerfield, IL).
Comment(s)	1	Studies used hospitalized patients and a computerized approach to removing outliers adapted from the Hoffmann technique. Values are the 97.5th percentiles.
	2	*Numbers not provided.
		Reference ranges were obtained by comparing results from previously published data and using regression equations.

UREA NITROGEN

Test	Age	Male			Female		
		n	mg/dL	mmol/L	n	mg/dL	mmol/L
1	1–7 d	171	2–13	0.7–4.6	114	2–13	0.7–4.6
	8–30 d	209	2–16	0.7–5.7	154	2–15	0.7–5.4
	1–3 mo	278	2–12	0.7–4.3	274	2–14	0.7–5.0
	4–6 mo	144	1–14	0.4–5.0	139	1–13	0.4–4.6
	7–12 mo	204	2–14	0.7–5.0	160	1–13	0.4–4.6
2	1–30 d	51	4–12	1.4–4.3	43	3–17	1.1–6.1
	1–12 mo	69	2–13	0.7–4.6	60	4–14	1.4–5.0
	1–3 y	104	3–12	1.1–4.3	127	3–14	1.1–5.0
	4–6 y	140	3–16	1.1–5.7	122	4–14	1.4–5.0
	7–9 y	124	4–16	1.4–5.7	121	4–16	1.4–5.7
	10–12 y	133	5–18	1.8–6.4	125	5–16	1.8–5.7
	13–15 y	141	7–18	2.5–6.4	153	4–15	1.4–5.4
	16–18 y	111	5–20	1.8–7.1	120	4–15	1.4–5.4

		Male and Female		
Test	Age	n	mg/dL	mmol/L
3	0–12 mo	130	5–14	1.8–4.9
	1–5 y	156	6–20	2.0–7.2
	6–10 y	134	7–19	2.4–6.8

	Male			Female		
Age	n	mg/dL	mmol/L	n	mg/dL	mmol/L
11–14 y	43	6–19	2.2–6.7	111	6–18	2.1–6.6
15–20 y	35	6–20	2.2–7.0	147	5–18	1.9–6.3

		Male and Female		
Test	Age	n	mg/dL	mmol/L
4	1–3 y	50	5–17	1.8–6.0
	4–6 y	38	7–17	2.5–6.0
	7–9 y	72	7–17	2.5–6.0
	10–11 y	62	7–17	2.5–6.0
	12–13 y	73	7–17	2.5–6.0
	14–15 y	91	8–21	2.9–7.5
	16–19 y	107	8–21	2.9–7.5
5	1–7 d	*	1–13	0.3–3.5
	8–30 d		1–16	0.3–4.3
	1–3 mo		1–12	0.3–3.2
	4–12 mo		1–14	0.3–3.8
	1–3 y		4–17	1.1–4.6
	4–13 y		6–17	1.6–4.6
	14–19 y		7–21	1.9–5.7

Specimen Type(s)	1–3,5	Plasma/serum
	4	Plasma
Reference(s)	1	Soldin SJ, Savwoir TV, Guo Y. Pediatric reference ranges for gamma-glutamyltransferase and urea nitrogen during the first year of life on the Vitros 500 analyzer. Clin Chem 1997;43:S199. (Abstract)
	2	Soldin SJ, Bailey J, Beatey J, et al. Pediatric reference ranges for blood urea nitrogen (BUN) on the Hitachi 747 analyzer. Clin Chem 1996;42:S307. (Abstract)
	3	Chan MK, Seiden-Long I, Aytekin M, et al. Canadian Laboratory Initiative on Reference Interval Database (CALIPER): pediatric reference intervals for an integrated clinical chemistry and immunoassay analyzer, Abbott ARCHITECT ci8200. Clin Biochem 2009;42:885–91.
	4	Lockitch G, Halsted AC, Albersheim S, et al. Age- and sex-specific pediatric reference intervals for biochemistry analytes as measured with the Ektachem-700 analyzer. Clin Chem 1988;34:1622–5. (Abstract)
	5	Ghoshal AK, Soldin SJ. Evaluation of the Dade Behring Dimension RxL: integrated chemistry system-pediatric reference ranges. Clin Chim Acta 2003;331:135–46.
Method(s)	1,4	Urease method, Vitros 500 & 700 (Ortho-Clinical Diagnostics, Raritan, NJ).
	2	Hitachi 747 with Boehringer Mannheim reagents (Boehringer Mannheim Diagnostics, Indianapolis, IN).
	3	Abbott Architect ci8200 (Abbott Diagnostics, Abbott Park, IL).
	5	The urea nitrogen method employs a urease/glutamate dehydrogenase-coupled enzymatic technique. Siemens Dimension RxL Analyzer (Siemens Healthcare Diagnostics, Deerfield, IL).
Comment(s)	1,2	Study used hospitalized patients and a computerized approach adapted from the Hoffmann technique to obtain 2.5–97.5th percentiles.
	3	1459 serum/plasma specimens from children attending select outpatient clinics were collected from the five age groups noted above. Values are 2.5–97.5th percentiles.
	4	Study used normal healthy children. Values are 2.5–97th percentiles.
	5	*Numbers not provided. Reference ranges were obtained by comparing results from previously published data and using regression equations.

URIC ACID

Test	Age	Male n	Male mg/dL	Male μmol/L	Female n	Female mg/dL	Female μmol/L
1	1–30 d	72	1.3–4.9	80–290	47	1.4–6.2	80–370
	1–3 mo	83	1.4–5.3	80–310	64	1.4–5.8	80–340
	4–6 mo	104	1.5–6.3	90–370	52	1.4–6.2	80–370
	7–12 mo	97	1.5–6.6	90–390	65	1.5–6.2	90–370
2	1–3 y	49*	1.8–5.0	105–300	49*	1.8–5.0	105–300
	4–6 y	38*	2.2–4.7	130–280	38*	2.2–4.7	130–280
	7–9 y	72*	2.0–5.0	120–295	72*	2.0–5.0	120–295
	10–11 y	28	2.3–5.4	135–320	34	3.0–4.7	180–280
	12–13 y	32	2.7–6.7	160–400	40	3.0–5.8	180–345
	14–15 y	39	2.4–7.8	140–465	50	3.0–5.8	180–345
	16–19 y	41	4.0–8.6	235–510	68	3.0–5.9	180–350
3	1–30 d	84	1.2–3.9	71–230	91	1.0–4.6	59–271
	31–365 d	138	1.2–5.6	71–330	110	1.1–5.4	65–319
	1–3 y	149	2.1–5.6	124–330	105	1.8–5.0	106–295
	4–6 y	120	1.8–5.5	106–325	91	2.0–5.1	118–301
	7–9 y	123	1.8–5.4	106–319	115	1.8–5.5	106–325
	10–12 y	121	2.2–5.8	130–342	92	2.5–5.9	148–348
	13–15 y	106	3.1–7.0	183–413	116	2.2–6.4	130–378
	16–18 y	82	2.1–7.6	124–448	111	2.4–6.6	142–389
4	1–30 d	**	1.2–4.9	71–291	**	1.3–6.2	77–369
	1–3 mo		1.3–5.3	77–315		1.3–5.8	77–345
	4–6 mo		1.4–6.4	83–381		1.3–6.2	77–369
	7–12 mo		1.4–6.7	83–399		1.4–6.2	83–369
	1–3 y		1.7–5.0	101–297		1.7–5.0	101–297
	4–6 y		2.2–4.7	131–280		2.2–4.7	131–280
	7–9 y		1.9–5.0	113–297		1.9–5.0	113–297
	10–11 y		2.3–5.4	137–321		3.0–4.7	178–280
	12–13 y		2.7–6.8	161–404		3.0–5.8	178–345
	14–15 y		2.4–7.9	143–470		3.0–5.8	178–345
	16–19 y		4.0–8.7	238–518		3.0–5.9	178–351

Specimen Type(s)	1,3,4	Plasma/serum
	2	Serum
Reference(s)	1	Soldin SJ, Savwoir TV, Guo Y. Pediatric reference ranges for lactate dehydrogenase and uric acid during the first year of life on the Vitros 500 analyzer. Clin Chem 1997;43:S199. (Abstract)
	2	Lockitch G, Halstead AC, Albersheim S, et al. Age- and sex-specific pediatric reference intervals for biochemistry analytes as measured with the Ektachem-700 analyzer. Clin Chem 1988;34:1622–5.
	3	Soldin SJ, Bailey J, Beatey J, et al. Pediatric reference ranges for uric acid. Clin Chem 1996;42:S308. (Abstract)
	4	Ghoshal AK, Soldin SJ. Evaluation of the Dade Behring Dimension RxL: integrated chemistry system-pediatric reference ranges. Clin Chim Acta 2003;331:135–46.
Method(s)	1, 2	Uricase on Vitros analyzer (Ortho-Clinical Diagnostics, Raritan, NJ).
	3	Hitachi 747 analyzer with Boehringer Mannheim reagents (Boehringer Mannheim Diagnostics, Indianapolis, IN).
	4	Measurement of uric acid by monitoring the loss of absorbance at 293 nm following uricase treatment is generally recognized as being more specific and less subject to interference than other indirect methods. Principles of procedure: Uric acid, which absorbs light at 293 nm, is converted by uricase to allantoin, which is nonabsorbing at 293 nm. The change in absorbance at 293 nm due to the disappearance of uric acid is directly proportional to the concentration of uric acid in the sample and is measured using a bichromatic (293,700 nm) endpoint technique. Siemens Dimension RxL Analyzer (Siemens Healthcare Diagnostics, Deerfield, IL).
Comment(s)	1, 3	Study used hospitalized patients and a computerized approach adapted from the Hoffmann technique to obtain the 2.5–97.5th percentiles.
	2	Study performed on healthy children. Results are 2.5–97.5th percentiles.
		*No significant differences were found for males and females. These ranges were therefore derived from combined data.
	4	**Numbers not provided.
		Reference ranges were obtained by comparing results from previously published data and using regression equations.

URINE VOLUME (24 h)

	Male and Female	
Age	mL	L
Newborn	50–300	0.05–0.30
Infant	350–550	0.35–0.55
Child	500–1000	0.50–1.00
Adolescent	700–1400	0.70–1.40
Thereafter	600–1800	0.60–1.80
Specimen Type(s)	Urine (24 h)	
Reference(s)	Behrman RE, ed. Nelson textbook of pediatrics, 14th ed. Philadelphia, PA: WB Saunders Company, 1992:1824.	
Method(s)	Not given.	
Comment(s)	Varies with intake and other factors.	

VANILLYLMANDELIC ACID (VMA; 4-HYDROXY-3-METHOXYMANDELIC ACID) (URINE)

			Male and Female				
Test	Age	n	mg/g creatinine	mmol/mol creatinine	n	mg/24h	µmol/24h
1	0–1 y	37	<18.8	<10.7	48	<2.3	<11.6
	2–4 y	49	<11.0	<6.3	34	<3.0	<15.1
	5–9 y	79	<8.3	<4.7	20	<3.5	<17.7
	10–19 y	55	<8.2	<4.7	40	<6.0	<30.3
	>19 y	56	<6.0	<3.4	56	<6.8	<34.3
2	0–3 mo	12	5.9–37.0	3.4–21.1			
	3–12 mo	28	8.4–43.8	4.8–25.0			
	1–2 y	15	7.9–23.0	4.5–13.1			
	2–5 y	22	2.9–23.0	1.7–13.1			
	5–10 y	21	5.8–18.7	3.3–10.7			
	10–15 y	13	1.6–10.6	0.9–6.1			
	>15 y	8	2.8–8.3	1.6–4.7			

Specimen Type(s)	1,2	Urine
Reference(s)	1	Soldin SJ, Hill JG. Liquid chromatographic analysis for urinary 4-hydroxy-3-methoxymandelic acid and 4-hydroxy-3-methoxyphenyl acetic acid and its use in investigation of neural crest tumors. Clin Chem 1981;27:502–3.
	2	Tuchman M, Morris CL, Ramnaraine ML, et al. Value of random urinary homovanillic acid and vanillylmandelic acid levels in the diagnosis and management of patients with neuroblastoma: comparison with 24-h urine collections. Pediatrics 1985;75:324–8.
Method(s)	1	HPLC with electrochemical detection.
	2	Capillary gas chromatography.
Comment(s)	1	Analysis performed on patients under investigation of hypertension or not suspected of having a neural crest tumor. All patients studied free of neoplasia. Results are 95th percentile.
	2	Values are 0–100th percentiles. Normal values were obtained from 93 pediatric patients in whom the diagnosis of neural crest tumors had been excluded.

VITAMIN A (RETINOL)

		Male and Female		
Test	Age	n	μmol/L	μg/dL
1	At birth	25	0.49–1.81	14–52
	Adult (mothers)	25	0.71–2.27	20–65
2	1–6 y	62*	0.7–1.5	20–43
	7–12 y	23*	0.9–1.7	20–49
	13–19 y	24*	0.9–2.5	26–72
3	Term neonates (>37 weeks)	41	0.63–1.75	18–50
	Preterm neonates (<36.6 weeks)	58	0.46–1.61	13–46

Specimen Type(s)	1,2	Serum
	3	Plasma/serum
Reference(s)	1	Jansson L, Nilsson B. Serum retinol and retinal-binding protein in mothers and infants at delivery. Biol Neonate 1983;43:269–71.
	2	Lockitch G, Halstead A, Wadsworth L, et al. Age- and sex-specific pediatric reference intervals for zinc, copper, selenium, iron, vitamins A and E, and related proteins. Clin Chem 1988;34:1625–8.
	3	Cardona-Pérez A, et al. Cord blood retinol and retinol binding protein in preterm and term neonates. Nutrition Research 1996;16:191–6.
Method(s)	1	Affinity chromatography and HPLC. See reference.
	2,3	HPLC. See reference for description.
Comment(s)	1	Results are mean ±2 SDs. Retinol studied in 25 healthy newborns and their mothers.
	2	The study population was healthy children. Non-parametric methods were used to determine the 0.025 and 0.975 fractiles.
		*No significant differences were found for males and females. These ranges were therefore derived from combined data.
	3	Healthy term and preterm neonates studied. Results are 2.5–97.5th percentiles.

VITAMIN B$_{12}$

Test	Age	Male		Female	
		n	pmol/L	n	pmol/L
1	0–1 y	127	216–891	94	168–1117
	2–3 y	142	195–897	133	307–892
	4–6 y	156	181–795	111	231–1038
	7–9 y	103	200–863	103	182–866
	10–12 y	105	135–803	94	145–752
	13–18 y	159	158–638	159	134–605
2	Newborn[a]	*	130–590	*	130–590
	After newborn period[a]		100–520		100–520
	Adult[b]		120–700		120–700

Specimen Type(s)	1	Plasma
	2	Serum
Reference(s)	1	Hicks JM, Cook J, Godwin ID, et al. Vitamin B$_{12}$ and folate: Pediatric reference ranges. Arch Pathol Lab Med 1993;117:704–6.
	2	[a]Behrman RE, Vaughan VC, eds. Nelson textbook of pediatrics. Philadelphia, PA: WB Saunders, 1983:1827–60.
		[b]Hall CA, Bardwell SA, Allen ES, et al. Variation in plasma folate levels among groups of healthy persons. Am J Clin Nutr 1975;28:854–7.
Method(s)	1	Radioimmunoassay Quantaphase B12 (Bio-Rad, Hercules, CA).
	2	See references for particulars of assays used.
Comment(s)	1	Study used hospitalized patients and a computerized approach adapted from the Hoffmann technique. Values are 2.5–97.5th percentiles.
	2	*Numbers not available.

25–HYDROXY VITAMIN D$_3$ (25 OH VIT D$_3$)

Test	Age	Male Summer n	Male Summer ng/mL	Male Winter n	Male Winter ng/mL	Female Summer n	Female Summer ng/mL	Female Winter n	Female Winter ng/mL
1	1–30 d	61	6.2–33.4	46	3.3–28.8	62	6.4–30.7	34	1.9–32.0
	31–365 d	76	18.6–36.9	71	7.4–53.3	87	16.0–48.2	64	11.6–36.7
	1–3 y	144	6.9–46.8	90	10.1–39.0	129	11.6–48.9	109	11.3–41.1
	4–12 y	171	4.6–37.4	151	6.3–36.0	161	2.8–36.7	59	8.4–29.9
	13–18 y	121	4.3–31.4	41	2.0–29.1	130	2.3–28.3	63	1.8–20.4
2	1–30 d	61	11–44	46	8–38	62	11–41	34	6–42
	1 mo–<1 y	76	26–48	71	13–68	87	23–62	64	18–48
	1–3 y	144	12–60	90	16–51	129	18–62	109	17–53
	4–12 y	171	9–49	151	11–47	161	7–48	59	14–40
	13–18 y	121	9–41	41	6–39	130	7–38	63	6–28

Note: Space does not permit the inclusion of nmol/L values above; however, to convert values from ng/mL to nmol/L, multiply the ng/mL value by 2.496.

Male and Female, All Seasons

Test	Age	n	ng/mL	nmol/L
3	0–<3 mo	131	5–42	12.5–104.8
	3–<6 mo	135	9–60	22.5–149.8
	6 mo–<12 mo	147	18–58	45.0–144.8
	1–<3 y	394	15–54	37.4–134.8
	3–<10 y	619	14–46	35.0–114.8
	10–<13 y	286	11–50	27.5–124.8
	13–<15 y	275	10–44	25.0–109.8
	15–<18 y	390	8–45	20.0–112.3
	>18 y	421	8–56	20.0–139.8

Age	Summer n	Summer ng/mL	Summer nmol/L	Winter n	Winter ng/mL	Winter nmol/L
0–<3 y	227	15–57	37.4–142.3	202	10–57	25.0–142.3
3–<10 y	195	17–52	42.4–129.8	137	11–38	27.5–94.8
10–<18 y	293	12–49	30.0–122.3	260	8–39	20.0–97.3
>18 y	134	12–50	30.0–124.8	122	7–44	17.5–109.8

Specimen Type(s)	1–3	Serum
Reference(s)	1	Soldin SJ, Hicks JM, Bailey J, et al. Pediatric reference ranges for 25 hydroxy vitamin D during the summer and winter. Clin Chem 1997;43:S200. (Abstract)
	2	Soldin SJ, Murthy J, Lauber B, et al. Pediatric reference ranges for 25-hydroxy vitamin D using DiaSorin kit. Clin Chem 1998;44:A14. (Abstract)
	3	Soldin OP, Sharma H, Husted L, Soldin SJ. Pediatric reference intervals for aldosterone, 17alpha-hydroxyprogesterone, dehydroepoandrosterone, testosterone and 25-hydroxy vitamin D_3 using tandem mass spectrometry. Clin Biochem 2009;42:823–7.
Method(s)	1	Radioimmunoassay. Nichols Institute Diagnostics Kit (Nichols Institue Diagnostics, Capistrano, CA).
	2	Radioimmunoassay, DiaSorin Kit (DiaSorin, Stillwater, MN).
	3	Samples were analyzed using isotope dilution liquid chromatography tandem mass spectrometry (LC/MS/MS).
Comment(s)	1,2	Study used hospitalized patients and a computerized approach adapted from the Hoffmann technique. Values are the 2.5–97.5th percentiles.
	3	Reference intervals were determined for neonates and children 0–18 years of age. The study was conducted using outpatient samples obtained between January 1, 2004, and June 30, 2008. Values are the 2.5–97.5th percentiles. The values given are present-day reference intervals. The clinically desired ranges are 30–100 ng/mL (74.9–249.6 nmol/L).

VITAMIN E (α-TOCOPHEROL)

Test	Age	n	μmol/L	μg/mL
	Male and Female			
1	1–6 y	62*	7–21	3.0–9.0
	7–12 y	23*	10–21	4.0–9.0
	13–19 y	24*	13–24	6.0–10.1
2	Premature	**	1–8	0.5–3.5
	Full term		2–8	1.0–3.5
	2–5 mo		5–14	2.0–6.0
	6–24 mo		8–19	3.5–8.0
	2–12 y		13–21	5.5–9.0

Specimen Type(s)	1	Serum
	2	Plasma/serum
Reference(s)	1	Lockitch G, Halstead A, Wadsworth L, et al. Age- and sex-specific pediatric reference intervals for zinc, copper, selenium, iron, vitamins A and E, and related proteins. Clin Chem 1988;34:1625–8.
	2	Meites S, ed. Pediatric clinical chemistry, 3rd ed. Washington, DC: AACC Press, 1989:295–6.
Method(s)	1	HPLC.
	2	Hansen LG, Warwick WJ. An improved assay method for vitamins A and E using fluorometry. Am J Clin Path 1978;70:922–3.
		McWhirter WR. Plasma tocopherol in infants and children. Acta Path Scand 1975;64:446–8.
Comment(s)	1	The study population was healthy children. Non-parametric methods were used to determine the 0.025 and 0.975 fractiles.
		*No significant differences were found for males and females. These ranges were therefore derived from combined data.
	2	**See references for numbers used in study.

ZINC

Test	Age	Male			Female		
		n	μg/dL	μmol/L	n	μg/dL	μmol/L
1	0–5 d	27*	65–140	9.9–21.4	27*	65–140	9.9–21.4
	1–5 y	77*	67–118	10.3–18.1	77*	67–118	10.3–18.1
	6–9 y	44*	77–107	11.8–16.4	44*	77–107	11.8–16.4
	10–14 y	36	76–101	11.6–15.4	23	79–118	12.1–18.0
	15–19 y	55	64–104	9.8–15.4	31	60–101	9.2–15.4
2	3–5 y	38**	71–149	10.9–22.8	38**	71–149	10.9–22.8
	6–8 y	43**	70–128	10.7–19.6	43**	70–128	10.7–19.6
	9–11 y	52**	64–124	9.8–19.0	52**	64–124	9.8–19.0
	12–13 y	24**	45–125	6.9–19.1	24**	45–125	6.9–19.1
	14–16 y	29**	42–125	6.4–19.1	29**	42–125	6.4–19.1

Test	Age	Male and Female		
		n	μg/dL	μmol/L
3	0–<0.5 y	13	26–141	4.0–21.6
	0.5–<1 y	18	29–131	4.5–20.1
	1–<2 y	15	31–120	4.8–18.4
	2–<4 y	23	29–115	4.4–17.6
	4–<6 y	19	48–119	7.4–18.2
	6–<10 y	25	48–129	7.3–19.7
	10–<14 y	21	25–148	3.9–22.7
	14–<18 y	17	46–130	7.1–19.9

Specimen Type(s)	1,2	Serum
	3	Plasma/serum
Reference(s)	1	Lockitch G, Halstead A, Wadsworth L, et al. Age- and sex-specific pediatric reference intervals for zinc, copper, selenium, iron, vitamins A and E, and related proteins. Clin Chem 1988;34:1625–8.
	2	Malvy DJ-M, Arnaud J, Burtschy, B, et al. Reference values for serum, zinc and selenium of French healthy children. Eur J Epidemiol 1993;9:155–61.
	3	Rükgauer M, Klein J, Kruse–Jarres JD. Reference values for the trace elements copper, manganese, selenium, and zinc in the serum/plasma of children, Adolescents and Adults. J. Trace Elements Med Biol 1997;11:92–8.
Method(s)	1	Atomic absorption spectrometry with deuterium background correction. Varian AA–1475 (Varian Canada, Inc., Georgetown, Canada).
	2	Perkin Elmer Atomic Absorption Spectrophotometry (Perkin Elmer Corporation, Rockville, MD).
	3	Atomic Absorption Spectrophotometry with Zeeman background compensation. Perkin Elmer ETAAS, Zeeman 3030 (Uberlingen, Germany).
Comment(s)	1	The study population was healthy children. Non-parametric methods were used to determine the 0.025 and 0.975 fractiles. *No significant differences were found for males and females. These ranges were therefore derived from combined data.
	2	The study population consisted of healthy French children. Values reported are the 2.5–97.5th percentiles. **No significant differences were found for males and females. These ranges were therefore derived from combined data.
	3	Study population was drawn from patients visiting the outpatient department or surgical or orthopedic ward for preoperative work up. Results are mean ±2 SDs (2.5th–97.5th percentiles).

ZINC PROTOPORPHYRIN

Age	Male			Female		
	n	µg/dL	µmol/mol heme	n	µg/dL	µmol/mol heme
0–12 mo	145	8.5–34.5	15.6–63.5	203	9–40	16.6–73.6
13–24 mo	695	10–34	18.4–62.6	725	11–32	20.2–58.9
2–5 y	1926	5–35	9.2–64.4	1822	10–31	18.4–57.0
6–9 y	522	6–31	11.0–57.0	408	9–30	16.6–55.2
10–17 y	61	5.5–31.5	10.1–58.0	61	5–33.5	9.2–61.6
Specimen Type(s)	50 µL anticoagulated whole blood containing either heparin or EDTA					
Reference(s)	Soldin OP, Miller M, Soldin SJ. Pediatric reference ranges for zinc protoporphyrin. Clin Biochem 2003;36:21–5.					
Method(s)	ZP Hematofluorometer Model 206 (AVIV Biomedical Inc., Lakewood, NJ).					
Comment(s)	Study used hospitalized patients and a computerized approach to removing outliers. Values are 2.5–97.5th percentiles.					

HEMATOLOGY TESTS

ATYPICAL LYMPHOCYTE COUNT (RELATIVE)

Age	Male		Female	
	n	%	n	%
≤3 d	196	2–2	155	2–2
4–7 d	123	2–4	146	2–4
8–14 d	139	2–4	177	2–3
15–30 d	206	2–4	191	2–3
31–60 d	290	2–3	243	2–3
61–180 d	812	2–3	658	2–4
0.5–<2 y	1731	2–4	1472	2–4
2–<6 y	2685	2–4	2348	2–4
6–<12 y	2890	2–3	2382	2–3
12–<18 y	3488	2–2	3755	2–2
≥18 y	1789	2–2	2622	2–2

Specimen Type(s)	Whole blood (K_2/EDTA anticoagulant)
Reference(s)	Brugnara C. Unpublished data.
Method(s)	Bayer ADVIA 120 (Bayer Diagnostics, Tarrytown, NY).
Comment(s)	Data from inpatients, excluding hematology/oncology patients. Data was analyzed using SAS statistical software. Outliers were removed according to the Chauvenet's Criterion (1). This criterion uses an exclusion factor based on the number of values in the sample set and is reapplied until no more outliers are excluded. The remaining data was truncated based on a method described by Hoffmann—the top and bottom 25% of data, based on an assumed normal distribution, was discarded (2). The method was modified to allow for computerized handling of the data. The 2.5th and 97.5th percentiles were calculated on the remaining 50% of the data. 1. Young HD. Statistical treatment of experimental data. New York: McGraw-Hill, 1962:76–9,162. 2. Hoffmann RG. Statistics in the practice of medicine. JAMA 1963;185:864–73.

BASOPHIL COUNT (ABSOLUTE)

Age	Male			Female		
	n	× 10³/μL	× 10⁹/L	n	× 10³/μL	× 10⁹/L
0–<14 d	101	0.02–0.11	0.02–0.11	72	0.02–0.07	0.02–0.07
15–30 d	143	0.01–0.07	0.01–0.07	93	0.01–0.06	0.01–0.06
31–60 d	189	0.01–0.07	0.01–0.07	161	0.01–0.05	0.01–0.05
61–<180 d	1005	0.01–0.06	0.01–0.06	493	0.01–0.07	0.01–0.07
0.5–<2 y	1840	0.01–0.06	0.01–0.06	1674	0.01–0.06	0.01–0.06
2–<6 y	2567	0.01–0.06	0.01–0.06	2072	0.01–0.06	0.01–0.06
6–<12 y	3154	0.01–0.06	0.01–0.06	2756	0.01–0.05	0.01–0.05
12–<18 y	4059	0.01–0.05	0.01–0.05	4935	0.01–0.05	0.01–0.05
≥18 y	975	0.01–0.05	0.01–0.05	1402	0.01–0.05	0.01–0.05

Specimen Type(s)	Whole blood (K$_3$/EDTA anticoagulant)
Reference(s)	Children's National Medical Center, Washington, DC. Unpublished data.
Method(s)	Sysmex XE-2100 (Sysmex Corporation, Kobe, Japan).
Comment(s)	Data from outpatients and ER patients, excluding hematology/oncology clinic patients. Data was log transformed and truncated based on a method described by Hoffmann. The top and bottom 30% of data, based on an assumed normal distribution, was discarded (1). The method was modified to allow for computerized handling of the data. The 2.5th and 97.5th percentile were calculated on the remaining 40% of data. 1. Hoffmann RG. Statistics in the practice of medicine. JAMA 1963;185:864–73.

BASOPHIL COUNT (RELATIVE)

Test	Age	Male			Female		
		n	%	Number Fraction	n	%	Number Fraction
1	≤3 d	118	0–1	.00–.01	97	0–1	.00–.01
	4–7 d	61	0–1	.00–.01	81	0–1	.00–.01
	8–14 d	85	1–1	.00–.01	89	0–1	.00–.01
	15–30 d	118	0–1	.00–.01	106	0–1	.00–.01
	31–60 d	162	0–1	.00–.01	161	0–1	.00–.01
	61–180 d	602	1–1	.00–.01	464	1–1	.00–.01
	0.5–<2 y	1283	1–1	.00–.01	1073	1–1	.00–.01
	2–<6 y	2340	0–1	.00–.01	1947	1–1	.00–.01
	6–<12 y	2732	0–1	.00–.01	2248	0–1	.00–.01
	12–<18 y	3306	0–1	.00–.01	3670	0–1	.00–.01
	≥18 y	1718	0–1	.00–.01	2573	0–1	.00–.01
2	0–14 d	100	0.1–0.8	.001–.008	76	0.1–0.6	.001–.006
	15–30 d	142	0.0–0.6	.000–.006	91	0.0–0.5	.000–.005
	31–60 d	192	0.0–0.6	.000–.006	165	0.0–0.5	.000–.005
	61–<180 d	518	0.0–0.6	.000–.006	496	0.0–0.6	.000–.006
	0.5–<2 y	1854	0.0–0.6	.000–.006	1690	0.0–0.6	.000–.006
	2–<6 y	2586	0.1–0.6	.001–.006	2092	0.0–0.6	.000–.006
	6–<12 y	3180	0.0–0.7	.000–.007	2777	0.0–0.6	.000–.006
	12–<18 y	4073	0.0–0.7	.000–.007	4953	0.0–0.6	.000–.006
	≥18 y	975	0.0–0.7	.000–.007	1405	0.0–0.7	.000–.007

Specimen Type(s)	1	Whole blood (K_2/EDTA anticoagulant)
	2	Whole blood (K_3/EDTA anticoagulant)
Reference(s)	1	Brugnara C. Unpublished data.
	2	Children's National Medical Center, Washington, DC. Unpublished data.
Method(s)	1	Bayer ADVIA 120 (Bayer Diagnostics, Tarrytown, NY).
	2	Sysmex XE-2100 (Sysmex Corporation, Kobe, Japan).
Comment(s)	1	Data from inpatients, excluding hematology/oncology patients. Data was analyzed using SAS statistical software. Outliers were removed according to the Chauvenet's Criterion (1). This criterion uses an exclusion factor based on the number of values in the sample set and is reapplied until no more outliers are excluded. The remaining data was truncated based on a method described by Hoffman—the top and bottom 25% of data, based on an assumed normal distribution, was discarded (2). The method was modified to allow for computerized handling of the data. The 2.5th and 97.5th percentiles were calculated on the remaining 50% of the data.
	2	Data from outpatients and ER patients, excluding hematology/oncology clinic patients. Data was truncated based on a method described by Hoffmann. The top and bottom 25% of data, based on an assumed normal distribution, was discarded (2). The method was modified to allow for computerized handling of the data. The 2.5th and 97.5th percentiles were calculated on the remaining 50% of the data.
		1. Young HD. Statistical treatment of experimental data. New York: McGraw-Hill, 1962:76–9,162.
		2. Hoffmann RG. Statistics in the practice of medicine. JAMA 1963;185:864–73.

CELLULAR HEMOGLOBIN CONCENTRATION DISTRIBUTION WIDTH (HDW)

Age	Male		Female	
	n	g/dL	n	g/dL
≤3 d	429	3.17–3.63	364	3.18–3.58
4–7 d	276	2.99–3.41	334	3.00–3.37
8–14 d	302	2.89–3.40	356	2.89–3.46
15–30 d	341	2.85–3.31	411	3.00–3.61
31–60 d	482	2.97–3.45	411	2.88–3.48
61–180 d	1185	2.84–3.35	1099	2.80–3.39
0.5–<2 y	2909	2.76–3.36	2377	2.69–3.17
2–<6 y	4067	2.75–3.21	3650	2.69–3.17
6–<12 y	4086	2.69–3.15	3299	2.63–3.06
12–<18 y	4994	2.65–3.15	5239	2.56–3.05
≥18 y	2496	2.64–3.19	3826	2.51–3.03

Specimen Type(s)	Whole blood (K$_2$/EDTA anticoagulant)
Reference(s)	Brugnara C. Unpublished data.
Method(s)	Bayer ADVIA 120 (Bayer Diagnostics, Tarrytown, NY).
Comment(s)	Data from inpatients, excluding hematology/oncology patients. Data was analyzed using SAS statistical software. Outliers were removed according to the Chauvenet's Criterion (1). This criterion uses an exclusion factor based on the number of values in the sample set and is reapplied until no more outliers are excluded. The remaining data was truncated based on a method described by Hoffmann—the top and bottom 25% of data, based on an assumed normal distribution, was discarded (2). The method was modified to allow for computerized handling of the data. The 2.5th and 97.5th percentiles were calculated on the remaining 50% of the data. 1. Young HD. Statistical treatment of experimental data. New York: McGraw-Hill, 1962:76–9,162. 2. Hoffmann RG. Statistics in the practice of medicine. JAMA 1963;185:864–73.

EOSINOPHIL COUNT (ABSOLUTE)

Test	Age	Male			Female		
		n	× 10³/μL	× 10⁹/L	n	× 10³/μL	× 10⁹/L
1	1–3 d	322	0.07–0.39	0.07–0.39	249	0.05–0.32	0.05–0.32
	4–7 d	175	0.09–0.42	0.09–0.42	209	0.05–0.39	0.05–0.39
	8–14 d	212	0.07–0.35	0.07–0.35	250	0.05–0.29	0.05–0.29
	15–30 d	277	0.12–0.51	0.12–0.51	302	0.03–0.37	0.03–0.37
	31–60 d	411	0.10–0.42	0.10–0.42	330	0.08–0.32	0.08–0.32
	61–180 d	997	0.05–0.36	0.05–0.36	883	0.01–0.28	0.01–0.28
	0.5–<2 y	2189	0.03–0.29	0.03–0.29	1759	0.01–0.20	0.01–0.20
	2–<6 y	3121	0.04–0.23	0.04–0.23	2755	0.02–0.20	0.02–0.20
	6–<12 y	3311	0.05–0.22	0.05–0.22	2690	0.04–0.19	0.04–0.19
	12–<18 y	4043	0.04–0.20	0.04–0.20	4192	0.05–0.17	0.05–0.17
	≥18 y	2090	0.06–0.23	0.06–0.23	2948	0.06–0.18	0.06–0.18
2	0–14 d	101	0.12–0.66	0.12–0.66	73	0.09–0.64	0.09–0.64
	15–30 d	143	0.08–0.80	0.08–0.80	93	0.06–0.75	0.06–0.75
	31–60 d	189	0.05–0.57	0.05–0.57	161	0.04–0.63	0.04–0.63
	61–<180 d	511	0.03–0.61	0.03–0.61	493	0.02–0.74	0.02–0.74
	0.5–<2 y	1840	0.02–0.82	0.02–0.82	1674	0.02–0.58	0.02–0.58
	2–<6 y	2567	0.03–0.53	0.03–0.53	2072	0.03–0.46	0.03–0.46
	6–<12 y	3154	0.03–0.52	0.03–0.52	2756	0.03–0.47	0.03–0.47
	12–<18 y	4059	0.04–0.38	0.04–0.38	4934	0.02–0.32	0.02–0.32
	≥18 y	956	0.03–0.44	0.03–0.44	1402	0.03–0.27	0.03–0.27

Specimen Type(s)	1	Whole blood (K_2/EDTA anticoagulant)
	2	Whole blood (K_3/EDTA anticoagulant)
Reference(s)	1	Brugnara C. Unpublished data.
	2	Children's National Medical Center, Washington, DC. Unpublished data.
Method(s)	1	Bayer ADVIA 120 (Bayer Diagnostics, Tarrytown, NY).
	2	Sysmex XE-2100 (Sysmex Corporation, Kobe, Japan).
Comment(s)	1	Data from inpatients, excluding hematology/oncology patients. Data was analyzed using SAS statistical software. Outliers were removed according to the Chauvenet's Criterion (1). This criterion uses an exclusion factor based on the number of values in the sample set and is reapplied until no more outliers are excluded. The remaining data was truncated based on a method described by Hoffmann—the top and bottom 25% of data, based on an assumed normal distribution, was discarded (2). The method was modified to allow for computerized handling of the data. The 2.5th and 97.5th percentiles were calculated on the remaining 50% of the data.
	2	Data from outpatients and ER patients, excluding hematology/oncology clinic patients. Data was log transformed and truncated based on a method described by Hoffmann. The top and bottom 30% of data, based on an assumed normal distribution, was discarded (2). The method was modified to allow for computerized handling of the data. The 2.5th and 97.5th percentiles were calculated on the remaining 40% of the data.
		1. Young HD. Statistical treatment of experimental data. New York: McGraw-Hill, 1962:76–9,162.
		2. Hoffmann RG. Statistics in the practice of medicine. JAMA 1963;185:864–73.

EOSINOPHIL COUNT (RELATIVE)

Test	Age	Male n	Male %	Male Number Fraction	Female n	Female %	Female Number Fraction
1	1–3 d	256	2–6	.020–.060	196	2–4	.020–.040
	4–7 d	141	2–6	.020–.060	163	2–5	.020–.050
	8–14 d	182	2–5	.020–.050	178	2–5	.020–.050
	15–30 d	227	2–7	.020–.070	221	2–4	.020–.040
	31–60 d	342	2–5	.020–.050	278	2–6	.020–.060
	61–180 d	823	2–6	.020–.060	676	2–5	.020–.050
	0.5–<2 y	1596	1–5	.010–.050	1193	1–4	.010–.040
	2–<6 y	2655	1–4	.010–.040	2194	1–4	.010–.040
	6–<12 y	2911	2–4	.020–.040	2349	1–4	.010–.040
	12–<18 y	3487	2–4	.020–.040	3766	1–3	.010–.030
	≥18 y	1865	2–4	.020–.040	2666	2–3	.020–.030
2	0–14 d	104	0.3–5.2	.003–.052	79	0.4–4.6	.004–.046
	15–30 d	148	0.2–5.4	.002–.054	98	0.0–5.3	.000–.053
	31–60 d	204	0.0–4.5	.000–.045	176	0.0–4.1	.000–.041
	61–<180 d	570	0.0–4.0	.000–.040	552	0.0–3.6	.000–.036
	0.5–<2 y	2029	0.0–3.7	.000–.037	1888	0.0–3.2	.000–.032
	2–<6 y	2864	0.0–4.1	.000–.041	2303	0.0–3.3	.000–.033
	6–<12 y	3512	0.0–4.7	.000–.047	3085	0.0–4.0	.000–.040
	12–<18 y	4503	0.0–4.0	.000–.040	5518	0.0–3.4	.000–.034
	≥18 y	1107	0.0–4.4	.000–.044	1619	0.0–3.0	.000–.030

Specimen Type(s)	1 Whole blood (K_2/EDTA anticoagulant)
	2 Whole blood (K_3/EDTA anticoagulant)
Reference(s)	1 Brugnara C. Unpublished data.
	2 Children's National Medical Center, Washington, DC. Unpublished data.
Method(s)	1 Bayer ADVIA 120 (Bayer Diagnostics, Tarrytown, NY).
	2 Sysmex XE-2100 (Sysmex Corporation, Kobe, Japan).
Comment(s)	1 Data from inpatients, excluding hematology/oncology patients. Data was analyzed using SAS statistical software. Outliers were removed according to the Chauvenet's Criterion (1). This criterion uses an exclusion factor based on the number of values in the sample set and is reapplied until no more outliers are excluded. The remaining data was truncated based on a method described by Hoffmann—the top and bottom 25% of data, based on an assumed normal distribution, was discarded (2). The method was modified to allow for computerized handling of the data. The 2.5th and 97.5th percentiles were calculated on the remaining 50% of the data.
	2 Data from outpatients and ER patients, excluding hematology/oncology clinic patients. Data was truncated based on a method described by Hoffmann. The top and bottom 30% of data, based on an assumed normal distribution, was discarded (2). The method was modified to allow for computerized handling of the data. The 2.5th and 97.5th percentiles were calculated on the remaining 40% of the data.
	1. Young HD. Statistical treatment of experimental data. New York: McGraw-Hill, 1962:76–9,162.
	2. Hoffmann RG. Statistics in the practice of medicine. JAMA 1963;185:864–73.

HEMATOCRIT

Test	Age	Male			Female		
		n	%	Volume Fraction	n	%	Volume Fraction
1	1–3 d	438	36.4–47.4	.364–.474	362	36.5–47.7	.365–.477
	4–7 d	276	35.9–46.6	.359–.466	333	36.1–44.0	.361–.440
	8–14 d	304	34.4–45.4	.344–.454	356	36.6–43.2	.366–.432
	15–30 d	346	33.6–41.0	.336–.410	418	34.1–41.8	.341–.418
	31–60 d	485	29.1–36.6	.291–.366	409	32.0–39.9	.320–.399
	61–180 d	1195	30.5–37.7	.305–.377	1113	30.5–38.6	.305–.386
	0.5–<2 y	2900	30.5–36.4	.305–.364	2381	30.9–36.4	.309–.364
	2–<6 y	4124	31.5–36.8	.315–.368	3712	31.8–37.0	.318–.370
	6–<12 y	4181	31.5–38.0	.315–.380	3366	32.3–38.3	.323–.383
	12–<18 y	5094	31.4–41.0	.314–.410	5351	32.1–38.7	.321–.387
	≥18 y	2527	33.0–43.4	.330–.434	3869	31.7–39.1	.317–.391
2	0–14 d	68	39.8–53.6	.398–.536	51	39.6–57.2	.396–.572
	15–30 d	58	30.5–45.0	.305–.450	42	32.0–44.5	.320–.445
	31–60 d	116	26.8–37.5	.268–.375	77	27.7–35.1	.277–.351
	61–180 d	305	28.6–37.2	.286–.372	231	29.5–37.1	.295–.371
	0.5–<2 y	1538	30.8–37.8	.308–.378	1343	30.9–37.9	.309–.379
	2–<6 y	1765	31.0–37.7	.310–.377	1680	31.2–37.8	.312–.378
	6–<12 y	1520	32.2–39.8	.322–.398	1304	32.4–39.5	.324–.395
	12–<18 y	1598	33.9–43.5	.339–.435	1944	33.4–40.4	.334–.404
	≥18 y	275	36.2–46.3	.362–.463	486	32.9–41.2	.329–.412

Specimen Type(s)	1 Whole blood (K_2/EDTA anticoagulant)
	2 Whole blood (K_3/EDTA anticoagulant)
Reference(s)	1 Brugnara C. Unpublished data.
	2 Children's National Medical Center, Washington, DC. Unpublished data.
Method(s)	1 Bayer ADVIA 120 (Bayer Diagnostics, Tarrytown, NY).
	2 Sysmex XE-2100 (Sysmex Corporation, Kobe, Japan).
Comment(s)	1 Data from inpatients, excluding hematology/oncology patients. Data was analyzed using SAS statistical software. Outliers were removed according to the Chauvenet's Criterion (1). This criterion uses an exclusion factor based on the number of values in the sample set and is reapplied until no more outliers are excluded. The remaining data was truncated based on a method described by Hoffmann—the top and bottom 25% of data, based on an assumed normal distribution, was discarded (2). The method was modified to allow for computerized handling of the data. The 2.5th and 97.5th percentiles were calculated on the remaining 50% of the data.
	2 Performed on outpatients and ER patients. Hematology/oncology clinic patients were excluded. Data was analyzed using SAS statistical software. Outliers were removed according to the Chauvenet's Criterion (1). This criterion uses an exclusion factor based on the number of values in the sample set and is reapplied until no more outliers are excluded. The remaining data was truncated based on a method described by Hoffmann—the top and bottom 10% of data, based on an assumed normal distribution, was discarded (2). The method was modified to allow for computerized handling of the data. The 2.5th and 97.5th percentiles were calculated on the remaining 80% of the data.
	1. Young HD. Statistical treatment of experimental data. New York: McGraw-Hill, 1962:76–9,162.
	2. Hoffmann RG. Statistics in the practice of medicine. JAMA 1963;185:864–73.

HEMATOPOIETIC CELL PROGENITOR (ABSOLUTE)

		Male and Female	
Age	n	× 10³/μL	× 10⁹/L
≤2 d	87	0.00–0.007	0.00–0.007
2–<14 d	100	0.00–0.008	0.00–0.008
14–30 d	98	0.00–0.008	0.00–0.008
31–90 d	86	0.00–0.005	0.00–0.005
91–180 d	87	0.00–0.001	0.00–0.001
0.5–<2 y	100	0.00–0.001	0.00–0.001
2–<6 y	91	0.00–0.001	0.00–0.001
6–<12 y	100	0.00–0.001	0.00–0.001
12–<18 y	100	0.00–0.000	0.00–0.000
≥18 y	91	0.00–0.000	0.00–0.000

Specimen Type(s)	Whole blood (K_3/EDTA anticoagulant)
Reference(s)	Children's National Medical Center, Washington, DC. Unpublished data.
Method(s)	Sysmex XE-2100 (Sysmex Corporation, Kobe, Japan).
Comment(s)	Data from outpatients and ER patients, excluding hematology/oncology clinic patients. Data was log transformed prior to truncation based on a method described by Hoffmann (1) in which the top and bottom 2.5% of data, based on an assumed normal distribution, was discarded. 1. Hoffmann RG. Statistics in the practice of medicine. JAMA 1963;185:864–73.

HEMOGLOBIN

Test	Age	Male			Female		
		n	g/dL	g/L	n	g/dL	g/L
1	1–3 d	437	12.5–16.6	125–166	362	12.7–16.4	127–164
	4–7 d	276	12.5–16.3	125–163	333	12.6–15.3	126–153
	8–14 d	304	11.9–15.7	119–157	356	12.7–14.9	127–149
	15–30 d	344	11.6–14.2	116–142	417	11.6–14.3	116–143
	31–60 d	486	10.2–12.7	102–127	407	11.1–13.7	111–137
	61–180 d	1197	10.5–13.0	105–130	1113	10.7–13.4	107–134
	0.5–<2 y	2916	10.4–12.5	104–125	2377	10.8–12.6	108–126
	2–<6 y	4123	11.0–12.8	110–128	3714	11.1–12.9	111–129
	6–<12 y	4183	11.0–13.3	110–133	3366	11.3–13.4	113–134
	12–<18 y	5095	11.0–14.3	110–143	5349	11.3–13.4	113–134
	≥18 y	2526	11.4–15.1	114–151	3874	10.9–13.4	109–134
2	1–14 d	60	13.9–19.1	139–191	48	13.4–20.0	134–200
	15–30 d	55	10.0–15.3	100–153	37	10.8–14.6	108–146
	31–60 d	110	8.9–12.7	89–127	71	9.2–11.4	92–114
	61–180 d	277	9.6–12.4	96–124	209	9.9–12.4	99–124
	0.5–<2 y	1350	10.1–12.5	101–125	1171	10.2–12.7	102–127
	2–<6 y	1383	10.2–12.7	102–127	1281	10.2–12.7	102–127
	6–<12 y	1420	10.7–13.4	107–134	1207	10.6–13.2	106–132
	12–<18 y	1569	11.0–14.5	110–145	1872	10.8–13.3	108–133
	≥18 y	268	11.9–15.4	119–154	471	10.6–13.5	106–135

Specimen Type(s)	1	Whole blood (K_2/EDTA anticoagulant)
	2	Whole blood (K_3/EDTA anticoagulant)
Reference(s)	1	Brugnara C. Unpublished data.
	2	Children's National Medical Center, Washington, DC. Unpublished data.
Method(s)	1	Bayer ADVIA 120 (Bayer Diagnostics, Tarrytown, NY).
	2	Sysmex XE-2100 (Sysmex Corporation, Kobe, Japan).
Comment(s)	1	Data from inpatients, excluding hematology/oncology patients. Data was analyzed using SAS statistical software. Outliers were removed according to the Chauvenet's Criterion (1). This criterion uses an exclusion factor based on the number of values in the sample set and is reapplied until no more outliers are excluded. The remaining data was truncated based on a method described by Hoffmann—the top and bottom 25% of data, based on an assumed normal distribution, was discarded (2). The method was modified to allow for computerized handling of the data. The 2.5th and 97.5th percentiles were calculated on the remaining 50% of the data.
	2	Performed on outpatients and ER patients. Hematology/oncology clinic patients were excluded. Data was analyzed using SAS statistical software. Outliers were removed according to the Chauvenet's Criterion (1). This criterion uses an exclusion factor based on the number of values in the sample set and is reapplied until no more outliers are excluded. The remaining data was truncated based on a method described by Hoffmann—the top and bottom 10% of data, based on an assumed normal distribution, was discarded (2). The method was modified to allow for computerized handling of the data. The 2.5th and 97.5th percentiles were calculated on the remaining 80% of the data.
	1.	Young HD. Statistical treatment of experimental data. New York: McGraw-Hill, 1962:76–9,162.
	2.	Hoffmann RG. Statistics in the practice of medicine. JAMA 1963;185:864–73.

HEMOGLOBIN A

	Male and Female	
Age	n	%
0–6 d	65	2.4–22.4
7–14 d	21	8.5–19.8
15–45 d	46	12.9–51.1
46 d–<3 mo	35	35.8–77.3
3–<6 mo	45	75.3–96.6
6–<9 mo	26	81.1–97.6
9–<15 mo	49	91.2–98.3
15 mo–<2 y	41	94.4–98.0
2–<6 y	100	95.7–98.0
6–<18 y	72	96.5–97.7
≥18 y	119	96.8–97.5

Specimen Type(s)	Whole blood (K_3/EDTA anticoagulant)
Reference(s)	Children's National Medical Center, Washington, DC. Publication pending.
Method(s)	Capillarys™ 2 Hemoglobin Assay (Sebia, Norcross, GA).
Comment(s)	Data from outpatients and ER patients excluding hematology/oncology clinic patients, previously transfused patients, and patients with other hemoglobinopathies. Data was truncated based on the method described by Hoffmann—the top and bottom 20–30% of the data, based on an assumed normal distribution was discarded in order to achieve R^2 greater than 93% (1). Exception: 15–45 d data was log transformed prior to truncation. The method was modified to allow for computerized handling of the data. The 2.5th and 97.5th percentiles were calculated on the remaining 40–60% of the data with the exception of 7–14 d data where the central 20% was used. 1. Hoffmann RG. Statistics in the practice of medicine. JAMA 1963:185:864–73.

HEMOGLOBIN A2		
Male and Female		
Age	n	%
0–6 d	65	0.0–0.0
7–14 d	21	0.0–1.0
15–45 d	46	0.0–1.5
46 d–<3 mo	35	0.6–1.9
3–<6 mo	45	1.7–2.8
6–<9 mo	26	2.1–2.9
9–<15 mo	49	2.2–3.2
15 mo–<2 y	41	2.3–3.1
2–<6 y	100	2.3–3.2
6–<18 y	72	2.4–3.0
≥18 y	119	2.4–3.0

Specimen Type(s)	Whole blood (K_3/EDTA anticoagulant)
Reference(s)	Children's National Medical Center, Washington, DC. Publication pending.
Method(s)	Capillarys™ 2 Hemoglobin Assay (Sebia, Norcross, GA).
Comment(s)	Data from outpatients and ER patients excluding hematology/oncology clinic patients, previously transfused patients, and patients with other hemoglobinopathies. Data was log transformed prior to truncation based on the method described by Hoffmann—the top and bottom 20–30% of the data, based on an assumed normal distribution was discarded in order to achieve R^2 greater than 95% (1). The method was modified to allow for computerized handling of the data. The 2.5th and 97.5th percentiles were calculated on the remaining 40–60% of the data with the exception of the 45 d–<3 mo and 3–<6 mo data where the log transformation was not performed. 1. Hoffmann RG. Statistics in the practice of medicine. JAMA 1963:185:864–73.

HEMOGLOBIN F

	Male and Female	
Age	n	%
0–6 d	65	77.0–97.9
7–14 d	21	79.6–91.4
15–45 d	46	59.8–89.6
46 d–<3 mo	35	23.9–67.2
3–<6 mo	45	4.4–27.5
6–<9 mo	26	1.5–27.8
9–<15 mo	49	0.4–8.4
15 mo–<2 y	41	0.1–4.9
2–<6 y	100	0.0–3.7
6–<17 y	72	0.0–1.3
≥18 y	119	0.0–0.0

Specimen Type(s)	Whole blood (K_3/EDTA anticoagulant)
Reference(s)	Children's National Medical Center, Washington, DC. Publication pending.
Method(s)	Capillarys™ 2 Hemoglobin Assay (Sebia, Norcross, GA).
Comment(s)	Data from outpatients and ER patients excluding hematology/oncology clinic patients, previously transfused patients, and patients with other hemoglobinopathies. Data was log transformed prior to truncation based on the method described by Hoffmann—the top and bottom 20–30% of the data, based on an assumed normal distribution was discarded in order to achieve R^2 greater than 95% (1). The method was modified to allow for computerized handling of the data. The 2.5th and 97.5th percentiles were calculated on the remaining 40–60% of the data with the exception of 7–14 d and 15–45 d data where the central 20% was used. 1. Hoffmann RG. Statistics in the practice of medicine. JAMA 1963:185:864–73.

IMMATURE GRANULOCYTES (ABSOLUTE)

Age	n	× $10^3/\mu L$	× $10^9/L$
≤2 d	87	0.00–0.28	0.00–0.28
2–<14 d	100	0.00–0.27	0.00–0.27
14–30 d	98	0.00–0.22	0.00–0.22
31–90 d	86	0.00–0.09	0.00–0.09
91–180 d	87	0.00–0.06	0.00–0.06
0.5–<2 y	100	0.00–0.14	0.00–0.14
2–<6 y	91	0.00–0.06	0.00–0.06
6–<12 y	100	0.00–0.04	0.00–0.04
12–<18 y	100	0.00–0.03	0.00–0.03
≥18 y	91	0.00–0.09	0.00–0.09

All values are for Male and Female combined.

Specimen Type(s)	Whole blood (K_3/EDTA anticoagulant)
Reference(s)	Children's National Medical Center, Washington, DC. Unpublished data.
Method(s)	Sysmex XE-2100 (Sysmex Corporation, Kobe, Japan).
Comment(s)	Data from outpatients and ER patients, excluding hematology/oncology clinic patients. Data was log transformed prior to truncation based on a method described by Hoffmann (1) in which the top and bottom 2.5% of data, based on an assumed normal distribution, was discarded. 1. Hoffmann RG. Statistics in the practice of medicine. JAMA 1963;185:864–73.

IMMATURE GRANULOCYTES (RELATIVE)

Male and Female

Age	n	%	Number Fraction
≤2 d	87	0.0–1.7	.000–.017
2–<14 d	100	0.0–1.9	.000–.019
14–30 d	98	0.0–1.3	.000–.013
31–90 d	86	0.0–0.9	.000–.009
91–180 d	87	0.0–0.5	.000–.005
0.5–<2 y	100	0.0–0.9	.000–.009
2–<6 y	91	0.0–0.8	.000–.008
6–<12 y	100	0.0–0.3	.000–.003
12–<18 y	100	0.0–0.3	.000–.003
≥18 y	91	0.0–0.6	.000–.006

Specimen Type(s)	Whole blood (K_3/EDTA anticoagulant)
Reference(s)	Children's National Medical Center, Washington, DC. Unpublished data.
Method(s)	Sysmex XE-2100 (Sysmex Corporation, Kobe, Japan).
Comment(s)	Data from outpatients and ER patients, excluding hematology/oncology clinic patients. Data was log transformed prior to truncation based on a method described by Hoffmann (1) in which the top and bottom 2.5% of data, based on an assumed normal distribution, was discarded. 1. Hoffmann RG. Statistics in the practice of medicine. JAMA 1963;185:864–73.

	IMMATURE PLATELET FRACTION (IPF)			
	Male		Female	
Age	n	%	n	%
0–180 d	70	2.0–6.8	51	1.3–6.8
6–<24 mo	54	1.4–3.8	53	1.4–4.5
2–<6 y	55	1.1–3.6	65	1.0–3.6
6–<12 y	56	1.0–4.9	54	1.0–4.7
12–<18 y	56	1.6–6.1	63	1.4–6.4
≥18 y	62	1.6–7.1	52	1.6–4.9
Specimen Type(s)	Whole blood (K_3/EDTA anticoagulant)			
Reference(s)	Children's National Medical Center, Washington, DC. Publication pending.			
Method(s)	Sysmex XE-2100 (Sysmex Corporation, Kobe, Japan).			
Comment(s)	Data from outpatients and ER patients excluding hematology/oncology clinic patient, previously transfused patients and patients with other hemoglobinopathies. Data was log transformed prior to truncation based on the method described by Hoffmann — the top and bottom 20% of the data, based on an assumed normal distribution was discarded (1). The method was modified to allow for computerized handling of the data. The 2.5th and 97.5th percentiles were calculated on the remaining 60% of the data with the exception of 6–<24 mo (male) and 6–<12 y (female) data where 40% was used. 1. Hoffmann RG. Statistics in the practice of medicine. JAMA 1963:185:864–73.			

IMMATURE RETICULOCYTE FRACTION (IRF)

		Male and Female	
Age	n	Percent	Fraction
1–3 d	55	30.5–35.1	.305–.351
4–30 d	46	14.5–24.6	.145–.246
31–60 d	24	19.1–28.9	.191–.289
61–180 d	33	13.4–23.3	.134–.233
0.5–<2 y	123	11.4–25.8	.114–.258
2–<6 y	123	8.4–21.7	.84–.217
6–<12 y	115	8.9–24.1	.89–.241
12–<18 y	152	9.0–18.7	.90–.187
≥18 y	30	9.3–17.4	.93–.174

Specimen Type(s)	Whole blood (K_3/EDTA anticoagulant)
Reference(s)	Children's National Medical Center, Washington, DC. Unpublished data.
Method(s)	Sysmex XE-2100 (Sysmex Corporation, Kobe, Japan).
Comment(s)	Performed on outpatients, nursery patients, and ER patients. Hematology/ oncology clinic patients were excluded. Ages >6 mo did not include ER patients. Data was analyzed using SAS statistical software. Outliers were removed according to the Chauvenet's Criterion (1). This criterion uses an exclusion factor based on the number of values in the sample set and is reapplied until no more outliers are excluded. The remaining data was truncated based on a method described by Hoffmann—the top and bottom 25% of data, based on an assumed normal distribution, was discarded (2). The method was modified to allow for computerized handling of the data. The 2.5th and 97.5th percentiles were calculated on the remaining 50% of the data. 1. Young HD. Statistical treatment of experimental data. New York: McGraw-Hill, 1962:76–9,162. 2. Hoffmann RG. Statistics in the practice of medicine. JAMA 1963;185:864–73.

		Male			Female		
		LYMPHOCYTE COUNT (ABSOLUTE)					
Test	Age	n	× 10³/μL	× 10⁹/L	n	× 10³/μL	× 10⁹/L
1	1–3 d	320	1.84–3.58	1.84–3.58	258	1.68–2.85	1.68–2.85
	4–7 d	181	1.53–4.09	1.53–4.09	209	1.17–3.45	1.17–3.45
	8–14 d	230	1.35–3.86	1.35–3.86	247	1.46–3.78	1.46–3.78
	15–30 d	289	1.68–5.25	1.68–5.25	318	1.65–5.04	1.65–5.04
	31–60 d	417	2.22–5.63	2.22–5.63	342	2.15–5.14	2.15–5.14
	61–180 d	1035	2.34–5.45	2.34–5.45	911	1.88–5.39	1.88–5.39
	0.5–<2 y	2266	2.32–5.49	2.32–5.49	1863	2.03–5.68	2.03–5.68
	2–<6 y	3217	1.33–3.47	1.33–3.47	2878	1.46–3.78	1.46–3.78
	6–<12 y	3455	1.23–2.69	1.23–2.69	2805	1.23–2.76	1.23–2.76
	12–<18 y	4171	1.03–2.18	1.03–2.18	4398	1.17–2.30	1.17–2.30
	≥18 y	2109	0.95–2.04	0.95–2.04	3023	1.14–2.28	1.14–2.28
2	0–14 d	107	2.07–7.53	2.07–7.53	73	1.75–8.00	1.75–8.00
	15–30 d	148	2.11–8.38	2.11–8.38	99	2.42–8.20	2.42–8.20
	31–60 d	200	2.47–7.95	2.47–7.95	171	2.29–9.14	2.29–9.14
	61–<180 d	564	2.45–8.89	2.45–8.89	548	2.14–8.99	2.14–8.99
	0.5–<2 y	2012	1.56–7.83	1.56–7.83	1860	1.52–8.09	1.52–8.09
	2–<6 y	2841	1.13–5.52	1.13–5.52	2281	1.25–5.77	1.25–5.77
	6–<12 y	3482	0.97–3.96	0.97–3.96	3062	1.16–4.28	1.16–4.28
	12–<18 y	4484	0.97–3.26	0.97–3.26	5499	1.16–3.33	1.16–3.33
	≥18 y	1106	0.85–3.00	0.85–3.00	1615	1.16–3.18	1.16–3.18

Specimen Type(s)	1	Whole blood (K_2/EDTA anticoagulant)
	2	Whole blood (K_3/EDTA anticoagulant)
Reference(s)	1	Brugnara C. Unpublished data.
	2	Children's National Medical Center, Washington, DC. Unpublished data.
Method(s)	1	Bayer ADVIA 120 (Bayer Diagnostics, Tarrytown, NY).
	2	Sysmex XE-2100 (Sysmex Corporation, Kobe, Japan).
Comment(s)	1	Data from inpatients, excluding hematology/oncology patients. Data was analyzed using SAS statistical software. Outliers were removed according to the Chauvenet's Criterion (1). This criterion uses an exclusion factor based on the number of values in the sample set and is reapplied until no more outliers are excluded. The remaining data was truncated based on a method described by Hoffmann—the top and bottom 25% of data, based on an assumed normal distribution, was discarded (2). The method was modified to allow for computerized handling of the data. The 2.5th and 97.5th percentiles were calculated on the remaining 50% of the data.
	2	Data from outpatients and ER patients, excluding hematology/oncology clinic patients. Data was truncated based on a method described by Hoffmann. The top and bottom 10% of data, based on an assumed normal distribution, was discarded (2). The method was modified to allow for computerized handling of the data. The 2.5th and 97.5th percentiles were calculated on the remaining 80% of the data.
		1. Young HD. Statistical treatment of experimental data. New York: McGraw-Hill, 1962:76–9,162.
		2. Hoffmann RG. Statistics in the practice of medicine. JAMA 1963;185:864–73.

LYMPHOCYTE COUNT (RELATIVE)

Test	Age	Male			Female		
		n	%	Number Fraction	n	%	Number Fraction
1	1–3 d	327	14–41	.14–.41	265	11–40	.11–.40
	4–7 d	182	14–46	.14–.46	211	10–45	.10–.45
	8–14 d	229	9–47	.09–.47	256	8–46	.08–.46
	15–30 d	289	9–56	.09–.56	319	8–57	.08–.57
	31–60 d	420	12–68	.12–.68	343	16–68	.16–.68
	61–180 d	1052	16–68	.16–.68	909	15–68	.15–.68
	0.5–<2 y	2265	15–67	.15–.67	1868	13–70	.13–.70
	2–<6 y	3215	11–54	.11–.54	2898	11–59	.11–.59
	6–<12 y	3439	8–45	.08–.45	2764	10–47	.10–.47
	12–<18 y	4125	8–41	.08–.41	4387	8–39	.08–.39
	≥18 y	2167	7–36	.07–.36	3051	10–38	.10–.38
2	0–14 d	98	33.7–67.6	.337–.676	74	24.9–68.5	.249–.685
	15–30 d	142	33.6–76.8	.336–.768	90	31.9–82.7	.319–.827
	31–60 d	193	42.5–85.7	.425–.857	164	37.8–86.7	.378–.867
	61–180 d	518	40.7–83.7	.407–.837	499	30.4–85.6	.304–.856
	0.5–<2 y	1856	26.0–79.6	.260–.796	1694	27.4–79.9	.274–.799
	2–<6 y	2587	18.4–66.6	.184–.666	2097	18.1–68.6	.181–.686
	6–<12 y	3190	15.5–56.6	.155–.566	2783	16.7–57.8	.167–.578
	12–<18 y	4073	16.4–52.7	.164–.527	4952	18.2–49.8	.182–.498
	≥18 y	976	12.2–47.1	.122–.471	1407	18.2–47.4	.182–.474

Specimen Type(s)	1	Whole blood (K_2/EDTA anticoagulant)
	2	Whole blood (K_3/EDTA anticoagulant)
Reference(s)	1	Brugnara C. Unpublished data.
	2	Children's National Medical Center, Washington, DC. Unpublished data.
Method(s)	1	Bayer ADVIA 120 (Bayer Diagnostics, Tarrytown, NY).
	2	Sysmex XE-2100 (Sysmex Corporation, Kobe, Japan).
Comment(s)	1	Data from inpatient, excluding hematology/oncology patients. Data was analyzed using SAS statistical software. Outliers were removed according to the Chauvenet's Criterion (1). This criterion uses an exclusion factor based on the number of values in the sample set and is reapplied until no more outliers are excluded. The remaining data was truncated based on a method described by Hoffmann—the top and bottom 25% of data, based on an assumed normal distribution, was discarded (2). The method was modified to allow for computerized handling of the data. The 2.5th and 97.5th percentiles were calculated on the remaining 50% of the data.
	2	Data from outpatients and ER patients, excluding hematology/oncology clinic patients. Data was truncated based on a method described by Hoffmann. The top and bottom 25% of data, based on an assumed normal distribution, was discarded (2). The method was modified to allow for computerized handling of the data. The 2.5th and 97.5th percentiles were calculated on the remaining 50% of the data.
		1. Young HD. Statistical treatment of experimental data. New York: McGraw-Hill, 1962:76–9,162.
		2. Hoffmann RG. Statistics in the practice of medicine. JAMA 1963;185:864–73.

MEAN CORPUSCULAR HEMOGLOBIN (MCH)

Test	Age	Male		Female	
		n	pg	n	pg
1	1–3 d	439	32.8–36.4	366	31.7–36.3
	4–7 d	275	30.9–33.4	335	30.6–32.3
	8–14 d	304	30.4–33.0	355	30.5–31.9
	15–30 d	347	30.6–32.6	408	30.5–32.0
	31–60 d	483	30.0–32.0	408	29.8–31.7
	61–180 d	1190	27.6–29.9	1106	28.5–30.4
	0.5–<2 y	2916	26.0–29.0	2374	26.5–29.3
	2–<6 y	4126	26.8–29.4	3712	27.0–29.6
	6–<12 y	4116	27.5–29.7	3313	27.8–30.0
	12–<18 y	5032	28.2–30.5	5242	28.4–30.7
	≥18 y	2437	28.9–31.4	3799	28.3–30.9
2	1–14 d	62	31.3–35.6	44	31.1–35.9
	15–30 d	58	29.9–34.1	39	30.4–35.3
	31–60 d	104	27.8–32.0	78	28.0–32.5
	61–180 d	292	24.4–28.9	219	24.4–29.5
	0.5–<2 y	1262	22.7–27.2	1086	23.2–27.5
	2–<6 y	1266	23.7–28.3	1132	23.7–28.6
	6–<12 y	1341	24.9–29.2	1146	24.8–29.5
	2–<18 y	1536	25.2–30.2	1848	24.8–30.2
	≥18 y	268	26.5–31.4	475	25.3–30.9

Specimen Type(s)	1	Whole blood (K_2/EDTA anticoagulant)
	2	Whole blood (K_3/EDTA anticoagulant)
Reference(s)	1	Brugnara C. Unpublished data.
	2	Children's National Medical Center, Washington, DC. Unpublished data.
Method(s)	1	Bayer ADVIA 120 (Bayer Diagnostics, Tarrytown, NY).
	2	Sysmex XE-2100 (Sysmex Corporation, Kobe, Japan).
Comment(s)	1	Data from inpatients, excluding hematology/oncology patients. Data was analyzed using SAS statistical software. Outliers were removed according to the Chauvenet's Criterion (1). This criterion uses an exclusion factor based on the number of values in the sample set and is reapplied until no more outliers are excluded. The remaining data was truncated based on a method described by Hoffmann—the top and bottom 25% of data, based on an assumed normal distribution, was discarded (2). The method was modified to allow for computerized handling of the data. The 2.5th and 97.5th percentiles were calculated on the remaining 50% of the data.
	2	Performed on outpatients and ER patients. Hematology/oncology clinic patients were excluded. Data was analyzed using SAS statistical software. Outliers were removed according to the Chauvenet's Criterion (1). This criterion uses an exclusion factor based on the number of values in the sample set and is reapplied until no more outliers are excluded. The remaining data was truncated based on a method described by Hoffmann—the top and bottom 10% of data, based on an assumed normal distribution, was discarded (2). The method was modified to allow for computerized handling of the data. The 2.5th and 97.5th percentiles were calculated on the remaining 80% of the data.
		1. Young HD. Statistical treatment of experimental data. New York: McGraw-Hill, 1962:76–9,162.
		2. Hoffmann RG. Statistics in the practice of medicine. JAMA 1963;185:864–73.

MEAN CORPUSCULAR HEMOGLOBIN CONCENTRATION (MCHC)

		Male			Female		
Test	Age	n	%	Concentration Fraction	n	%	Concentration Fraction
1	1–3 d	440	34.0–35.3	.340–.353	365	33.9–35.4	.339–.354
	4–7 d	276	34.3–35.7	.343–.357	335	34.3–35.7	.343–.357
	8–14 d	303	34.0–35.6	.340–.356	358	33.9–35.3	.339–.353
	15–30 d	347	33.9–35.3	.339–.353	418	33.7–35.1	.337–.351
	31–60 d	485	34.0–35.5	.340–.355	409	34.1–35.4	.341–.354
	61–180 d	1192	33.9–35.4	.339–.354	1104	34.1–35.6	.341–.356
	0.5–<2 y	2925	33.6–35.2	.336–.352	2406	34.1–35.6	.341–.356
	2–<6 y	4138	34.2–35.7	.342–.357	3722	34.0–35.6	.340–.356
	6–<12 y	4137	34.4–35.8	.344–.358	3362	34.3–35.8	.343–.358
	12–<18 y	5042	34.2–35.6	.342–.356	5327	33.9–35.4	.339–.354
	≥18 y	2511	33.7–35.3	.337–.353	3876	33.6–35.0	.336–.350
2	0–14 d	64	33.0–35.7	.330–.357	49	33.4–35.4	.334–.354
	15–30 d	59	32.7–35.1	.327–.351	39	33.2–35.0	.332–.350
	31–60 d	116	32.3–34.8	.323–.348	76	32.5–34.9	.325–.349
	61–180 d	301	31.9–34.4	.319–.344	227	32.1–34.4	.321–.344
	0.5–<2 y	1260	31.6–34.4	.316–.344	1088	31.9–34.2	.319–.342
	2–<6 y	1273	32.0–34.7	.320–.347	1143	31.8–34.6	.318–.346
	6–<12 y	1360	32.2–34.9	.322–.349	1148	31.8–34.6	.318–.346
	12–<18 y	1553	31.8–34.8	.318–.348	1845	31.5–34.2	.315–.342
	≥18 y	271	31.9–34.8	.319–.348	483	31.0–34.1	.310–.341

Specimen Type(s)	1	Whole blood (K_2/EDTA anticoagulant)
	2	Whole blood (K_3/EDTA anticoagulant)
Reference(s)	1	Brugnara C. Unpublished data.
	2	Children's National Medical Center, Washington, DC. Unpublished data.
Method(s)	1	Bayer ADVIA 120 (Bayer Diagnostics, Tarrytown, NY).
	2	Sysmex XE-2100 (Sysmex Corporation, Kobe, Japan).
Comment(s)	1	Data from inpatients, excluding hematology/oncology patients. Data was analyzed using SAS statistical software. Outliers were removed according to the Chauvenet's Criterion (1). This criterion uses an exclusion factor based on the number of values in the sample set and is reapplied until no more outliers are excluded. The remaining data was truncated based on a method described by Hoffmann—the top and bottom 25% of data, based on an assumed normal distribution, was discarded (2). The method was modified to allow for computerized handling of the data. The 2.5th and 97.5th percentiles were calculated on the remaining 50% of the data.
	2	Performed on outpatients and ER patients. Hematology/oncology clinic patients were excluded. Data was analyzed using SAS statistical software. Outliers were removed according to the Chauvenet's Criterion (1). This criterion uses an exclusion factor based on the number of values in the sample set and is reapplied until no more outliers are excluded. The remaining data was truncated based on a method described by Hoffmann—the top and bottom 10% of data, based on an assumed normal distribution, was discarded (2). The method was modified to allow for computerized handling of the data. The 2.5th and 97.5th percentiles were calculated on the remaining 80% of the data.
		1. Young HD. Statistical treatment of experimental data. New York: McGraw-Hill, 1962:76–9,162.
		2. Hoffmann RG. Statistics in the practice of medicine. JAMA 1963;185:864–73.

MEAN CORPUSCULAR VOLUME (MCV)

Test	Age	Male n	Male μm³	Male fL	Female n	Female μm³	Female fL
1	1–3 d	439	94.0–106.3	94.0–106.3	366	89.7–105.4	89.7–105.4
	4–7 d	276	87.1–96.5	87.1–96.5	334	86.5–93.8	86.5–93.8
	8–14 d	304	87.1–94.8	87.1–94.8	350	87.4–92.2	87.4–92.2
	15–30 d	348	88.0–95.2	88.0–95.2	415	88.4–93.3	88.4–93.3
	31–60 d	483	86.5–92.1	86.5–92.1	410	85.7–91.6	85.7–91.6
	61–180 d	1192	79.6–86.3	79.6–86.3	1103	82.0–87.0	82.0–87.0
	0.5–<2 y	2898	75.6–83.1	75.6–83.1	2366	76.6–83.2	76.6–83.2
	2–<6 y	4126	76.8–83.3	76.8–83.3	3698	77.7–84.1	77.7–84.1
	6–<12 y	4097	78.2–83.9	78.2–83.9	3317	79.5–85.2	79.5–85.2
	12–<18 y	5046	80.8–86.6	80.8–86.6	5257	82.1–87.7	82.1–87.7
	≥18 y	2477	83.5–90.2	83.5–90.2	3815	82.7–89.4	82.7–89.4
2	0–14 d	63	91.3–103.1	91.3–103.1	48	92.7–106.4	92.7–106.4
	15–30 d	56	89.4–99.7	89.4–99.7	39	90.1–103.0	90.1–103.0
	31–60 d	110	84.3–94.2	84.3–94.2	78	83.4–96.4	83.4–96.4
	61–180 d	298	74.1–87.5	74.1–87.5	229	74.8–88.3	74.8–88.3
	0.5–<2 y	1260	69.5–81.7	69.5–81.7	1081	71.3–82.6	71.3–82.6
	2–<6 y	1279	71.3–84.0	71.3–84.0	1133	72.3–85.0	72.3–85.0
	6–<12 y	1349	74.4–86.1	74.4–86.1	1140	75.9–87.6	75.9–87.6
	12–<18 y	1543	76.7–89.2	76.7–89.2	1840	76.9–90.6	76.9–90.6
	≥18 y	273	80.0–93.6	80.0–93.6	477	77.7–93.7	77.7–93.7

Specimen Type(s)	1	Whole blood (K_2/EDTA anticoagulant)
	2	Whole blood (K_3/EDTA anticoagulant)
Reference(s)	1	Brugnara C. Unpublished data.
	2	Children's National Medical Center, Washington, DC. Unpublished data.
Method(s)	1	Bayer ADVIA 120 (Bayer Diagnostics, Tarrytown, NY).
	2	Sysmex XE-2100 (Sysmex Corporation, Kobe, Japan).
Comment(s)	1	Data from inpatients, excluding hematology/oncology patients. Data was analyzed using SAS statistical software. Outliers were removed according to the Chauvenet's Criterion (1). This criterion uses an exclusion factor based on the number of values in the sample set and is reapplied until no more outliers are excluded. The remaining data was truncated based on a method described by Hoffmann—the top and bottom 25% of data, based on an assumed normal distribution, was discarded (2). The method was modified to allow for computerized handling of the data. The 2.5th and 97.5th percentiles were calculated on the remaining 50% of the data.
	2	Performed on outpatients and ER patients. Hematology/oncology clinic patients were excluded. Data was analyzed using SAS statistical software. Outliers were removed according to the Chauvenet's Criterion (1). This criterion uses an exclusion factor based on the number of values in the sample set and is reapplied until no more outliers are excluded. The remaining data was truncated based on a method described by Hoffmann—the top and bottom 10% of data, based on an assumed normal distribution, was discarded (2). The method was modified to allow for computerized handling of the data. The 2.5th and 97.5th percentiles were calculated on the remaining 80% of the data.
		1. Young HD. Statistical treatment of experimental data. New York: McGraw-Hill, 1962:76–9,162.
		2. Hoffmann RG. Statistics in the practice of medicine. JAMA 1963;185:864–73.

MEAN PLATELET VOLUME (MPV)

Test	Age	Male n	Male μm³	Male fL	Female n	Female μm³	Female fL
1	1–3 d	430	7.8–8.5	7.8–8.5	344	7.9–8.5	7.9–8.5
	4–7 d	268	8.0–8.9	8.0–8.9	331	8.2–9.1	8.2–9.1
	8–14 d	291	8.1–9.1	8.1–9.1	350	8.3–9.4	8.3–9.4
	15–30 d	333	8.0–9.3	8.0–9.3	412	8.4–9.9	8.4–9.9
	31–60 d	472	7.8–8.9	7.8–8.9	390	7.8–8.8	7.8–8.8
	61–180 d	1157	7.5–8.3	7.5–8.3	1072	7.5–8.3	7.5–8.3
	0.5–<2 y	2674	7.3–8.1	7.3–8.1	2360	7.3–8.0	7.3–8.0
	2–<6 y	4065	7.2–7.9	7.2–7.9	3572	7.3–8.0	7.3–8.0
	6–<12 y	4106	7.4–8.1	7.4–8.1	3320	7.4–8.1	7.4–8.1
	12–<18 y	4981	7.5–8.3	7.5–8.3	5173	7.5–8.3	7.5–8.3
	≥18 y	2480	7.7–8.7	7.7–8.7	3837	7.8–8.8	7.8–8.8
2	0–14 d	57	10.2–11.9	10.2–11.9	41	10.4–12.0	10.4–12.0
	15–30 d	51	10.1–12.1	10.1–12.1	38	10.0–12.2	10.0–12.2
	31–60 d	100	9.2–10.8	9.2–10.8	69	9.4–11.1	9.4–11.1
	61–180 d	264	8.9–10.6	8.9–10.6	204	9.0–10.9	9.0–10.9
	0.5–<2 y	1100	8.7–10.5	8.7–10.5	981	8.8–10.6	8.8–10.6
	2–<6 y	1138	9.0–10.9	9.0–10.9	999	8.9–11.0	8.9–11.0
	6–<12 y	1248	9.2–11.4	9.2–11.4	1057	9.3–11.3	9.3–11.3
	12–<18 y	1403	9.6–11.8	9.6–11.8	1705	9.6–11.7	9.6–11.7
	≥18 y	240	9.7–11.9	9.7–11.9	431	9.6–12.0	9.6–12.0

Specimen Type(s)	1	Whole blood (K_2/EDTA anticoagulant)
	2	Whole blood (K_3/EDTA anticoagulant)
Reference(s)	1	Brugnara C. Unpublished data.
	2	Children's National Medical Center, Washington, DC. Unpublished data.
Method(s)	1	Bayer ADVIA 120 (Bayer Diagnostics, Tarrytown, NY).
	2	Sysmex XE-2100 (Sysmex Corporation, Kobe, Japan).
Comment(s)	1	Data from inpatients, excluding hematology/oncology patients. Data was analyzed using SAS statistical software. Outliers were removed according to the Chauvenet's Criterion (1). This criterion uses an exclusion factor based on the number of values in the sample set and is reapplied until no more outliers are excluded. The remaining data was truncated based on a method described by Hoffmann—the top and bottom 25% of data, based on an assumed normal distribution, was discarded (2). The method was modified to allow for computerized handling of the data. The 2.5th and 97.5th percentiles were calculated on the remaining 50% of the data.
	2	Performed on the XE-2100 on outpatients and ER patients. Hematology oncology clinic patients were excluded. Data was analyzed using SAS statistical software. Outliers were removed according to Chauvenet's Criterion (1). This criterion uses an exclusion factor based on the number of values in the sample set and is reapplied until no more outliers are excluded. The remaining data was truncated based on a method described by Hoffmann—the top and bottom 10% of data, based on an assumed normal distribution, was discarded (2). The method was modified to allow for computerized handling of the data. The 2.5th and 97.5th percentiles were calculated on the remaining 80% of the data.
		1. Young HD. Statistical treatment of experimental data. New York: McGraw-Hill, 1962:76–9,162.
		2. Hoffmann RG. Statistics in the practice of medicine. JAMA 1963;185:864–73.

MONOCYTE COUNT (ABSOLUTE)

Age	Male			Female		
	n	$\times 10^3/\mu L$	$\times 10^9/L$	n	$\times 10^3/\mu L$	$\times 10^9/L$
0–14 d	101	0.52–1.77	0.52–1.77	72	0.57–1.72	0.57–1.72
15–30 d	143	0.28–1.38	0.28–1.38	93	0.42–1.21	0.42–1.21
31–60 d	189	0.28–1.05	0.28–1.05	161	0.28–1.21	0.28–1.21
61–<180 d	511	0.28–1.07	0.28–1.07	493	0.24–1.17	0.24–1.17
0.5–<2 y	1840	0.25–1.15	0.25–1.15	1674	0.26–1.08	0.26–1.08
2–<6 y	2567	0.19–0.94	0.19–0.94	2072	0.24–0.92	0.24–0.92
6–<12 y	3153	0.19–0.85	0.19–0.85	2756	0.19–0.81	0.19–0.81
12–<18 y	4059	0.18–0.78	0.18–0.78	4934	0.19–0.72	0.19–0.72
≥18 y	960	0.19–0.77	0.19–0.77	1402	0.29–0.71	0.29–0.71

Specimen Type(s)	Whole blood (K$_3$/EDTA anticoagulant)
Reference(s)	Children's National Medical Center, Washington, DC. Unpublished data.
Method(s)	Sysmex XE-2100 (Sysmex Corporation, Kobe, Japan).
Comment(s)	Data from outpatients and ER patients, excluding hematology/oncology clinic patients. Data was truncated based on a method described by Hoffmann. The top and bottom 25% of data, based on an assumed normal distribution, was discarded (1). The method was modified to allow for computerized handling of the data. The 2.5th and 97.5th percentiles were calculated on the remaining 50% of the data. 1. Hoffmann RG. Statistics in the practice of medicine. JAMA 1963;185:864–73.

MONOCYTE COUNT (RELATIVE)

Test	Age	Male			Female		
		n	%	Number Fraction	n	%	Number Fraction
1	1–3 d	323	4–13	.04–.13	253	5–11	.05–.11
	4–7 d	177	7–17	.07–.17	200	6–14	.06–.14
	8–14 d	224	7–18	.07–.18	251	6–19	.06–.19
	15–30 d	281	6–18	.06–.18	310	5–14	.05–.14
	31–60 d	412	6–17	.06–.17	330	5–14	.05–.14
	61–180 d	1027	4–11	.04–.11	874	4–12	.04–.12
	0.5–<2 y	2180	4–10	.04–.10	1763	4–9	.04–.09
	2–<6 y	3034	4–9	.04–.09	2613	4–8	.04–.08
	6–<12 y	3297	4–8	.04–.08	2648	4–7	.04–.07
	12–<18 y	3959	4–8	.04–.08	4208	4–7	.04–.07
	≥18 y	2107	4–8	.04–.08	2942	4–7	.04–.07
2	0–14 d	98	6.7–19.9	.067–.199	77	5.2–20.6	.052–.206
	15–30 d	144	4.3–18.3	.043–.183	91	5.6–13.8	.056–.138
	31–60 d	193	4.4–14.0	.044–.140	166	3.8–15.5	.038–.155
	61–<180 d	518	3.8–13.4	.038–.134	497	3.8–12.6	.038–.126
	0.5–<2 y	1856	4.4–13.4	.044–.134	1696	3.8–12.8	.038–.128
	2–<6 y	2589	4.2–12.2	.042–.122	2097	4.1–11.4	.041–.114
	6–<12 y	3193	4.2–12.3	.042–.123	2783	4.2–11.3	.042–.113
	12–<18 y	4074	4.4–12.3	.044–.123	4960	4.1–10.9	.041–.109
	≥18 y	976	4.4–12.3	.044–.123	1405	4.3–11.0	.043–.110

Specimen Type(s)	1	Whole blood (K_2/EDTA anticoagulant)
	2	Whole blood (K_3/EDTA anticoagulant)
Reference(s)	1	Brugnara C. Unpublished data.
	2	Children's National Medical Center, Washington, DC. Unpublished data.
Method(s)	1	Bayer ADVIA 120 (Bayer Diagnostics, Tarrytown, NY).
	2	Sysmex XE-2100 (Sysmex Corporation, Kobe, Japan).
Comment(s)	1	Data from inpatients, excluding hematology/oncology patients. Data was analyzed using SAS statistical software. Outliers were removed according to the Chauvenet's Criterion (1). This criterion uses an exclusion factor based on the number of values in the sample set and is reapplied until no more outliers are excluded. The remaining data was truncated based on a method described by Hoffmann—the top and bottom 25% of data, based on an assumed normal distribution, was discarded (2). The method was modified to allow for computerized handling of the data. The 2.5th and 97.5th percentiles were calculated on the remaining 50% of the data.
	2	Data from outpatients and ER patients, excluding hematology/oncology clinic patients. Data was log transformed and truncated based on a method described by Hoffmann. The top and bottom 25% of data, based on an assumed normal distribution, was discarded (2). The method was modified to allow for computerized handling of the data. The 2.5th and 97.5th percentiles were calculated on the remaining 50% of the data.
		1. Young HD. Statistical treatment of experimental data. New York: McGraw-Hill, 1962:76–9,162.
		2. Hoffmann RG. Statistics in the practice of medicine. JAMA 1963;185:864–73.

NEUTROPHIL COUNT (ABSOLUTE)

Test	Age	Male			Female		
		n	× 10^3/μL	× 10^9/L	n	× 10^3/μL	× 10^9/L
1	1–3 d	311	4.33–9.11	4.33–9.11	260	4.43–11.43	4.43–11.43
	4–7 d	174	3.33–6.58	3.33–6.58	197	3.18–7.19	3.18–7.19
	8–14 d	223	3.71–8.36	3.71–8.36	235	3.91–8.26	3.91–8.26
	15–30 d	286	3.47–9.42	3.47–9.42	300	3.77–9.43	3.77–9.43
	31–60 d	406	2.20–6.45	2.20–6.45	337	2.49–6.26	2.49–6.26
	61–180 d	1035	2.57–7.54	2.57–7.54	906	2.22–7.11	2.22–7.11
	0.5–<2 y	2153	2.47–6.41	2.47–6.41	1817	2.34–6.44	2.34–6.44
	2–<6 y	3139	2.49–5.96	2.49–5.96	2764	2.29–6.36	2.29–6.36
	6–<12 y	3317	2.77–6.34	2.77–6.34	2733	2.58–5.95	2.58–5.95
	12–<18 y	4120	2.73–6.68	2.73–6.68	4271	3.04–6.06	3.04–6.06
	≥18 y	2116	3.54–7.52	3.54–7.52	2932	3.32–6.30	3.32–6.30
2	0–14 d	104	1.60–6.06	1.60–6.06	73	1.73–6.75	1.73–6.75
	15–30 d	147	1.18–5.45	1.18–5.45	99	1.23–4.80	1.23–4.80
	31–60 d	200	0.83–4.23	0.83–4.23	171	1.00–4.68	1.00–4.68
	61–<180 d	564	0.97–5.45	0.97–5.45	548	1.04–7.20	1.04–7.20
	0.5–<2 y	2012	1.19–7.21	1.19–7.21	1870	1.27–7.18	1.27–7.18
	2–<6 y	2842	1.54–7.92	1.54–7.92	2281	1.60–8.29	1.60–8.29
	6–<12 y	3497	1.63–7.55	1.63–7.55	3065	1.64–7.87	1.64–7.87
	12–<18 y	4487	1.54–7.04	1.54–7.04	5506	1.82–7.47	1.82–7.47
	≥18 y	1100	1.82–7.42	1.82–7.42	1615	2.00–7.15	2.00–7.15

Specimen Type(s)	1 Whole blood (K_2/EDTA anticoagulant)
	2 Whole blood (K_3/EDTA anticoagulant)
Reference(s)	1 Brugnara C. Unpublished data.
	2 Children's National Medical Center, Washington, DC. Unpublished data.
Method(s)	1 Bayer ADVIA 120 (Bayer Diagnostics, Tarrytown, NY).
	2 Sysmex XE-2100 (Sysmex Corporation, Kobe, Japan).
Comment(s)	1 Data from inpatients, excluding hematology/oncology patients. Data was analyzed using SAS statistical software. Outliers were removed according to the Chauvenet's Criterion (1). This criterion uses an exclusion factor based on the number of values in the sample set and is reapplied until no more outliers are excluded. The remaining data was truncated based on a method described by Hoffmann—the top and bottom 25% of data, based on an assumed normal distribution, was discarded (2). The method was modified to allow for computerized handling of the data. The 2.5th and 97.5th percentiles were calculated on the remaining 50% of the data.
	2 Data from outpatients and ER patients, excluding hematology/oncology clinic patients. Data was truncated based on a method described by Hoffmann. The top and bottom 25% of data, based on an assumed normal distribution, was discarded (2). The method was modified to allow for computerized handling of the data. The 2.5th and 97.5th percentiles were calculated on the remaining 50% of the data.
	1. Young HD. Statistical treatment of experimental data. New York: McGraw-Hill, 1962:76–9,162.
	2. Hoffmann RG. Statistics in the practice of medicine. JAMA 1963;185:864–73.

NEUTROPHIL COUNT (RELATIVE)

Test	Age	Male			Female		
		n	%	Number Fraction	n	%	Number Fraction
1	1–14 d	38	24.1–47.1	.241–.471	27	21.2–55.4	.212–.554
	15–30 d	37	18.4–32.4	.184–.324	29	17.0–40.9	.170–.409
	31–60 d	148	14.6–40.9	.146–.409	111	15.7–49.1	.157–.491
	61–180 d	472	17.0–55.5	.170–.555	361	18.6–60.0	.186–.600
	0.5–<2 y	1571	22.7–69.2	.227–.692	1272	23.8–69.3	.238–.693
	2–<6 y	1651	31.7–75.4	.317–.754	1328	33.6–77.5	.336–.775
	6–<12 y	1503	38.8–76.7	.388–.767	1406	38.7–76.7	.387–.767
	12–<18 y	1433	43.2–76.7	.432–.767	1786	46.4–75.6	.464–.756
	≥18 y	716	47.2–77.6	.472–.776	1094	50.4–77.7	.504–.777
2	0–14 d	54	20.2–46.2	.202–.462	53	15.2–66.1	.152–.661
	15–30 d	80	14.0–54.6	.140–.546	61	10.6–57.3	.106–.573
	31–60 d	69	10.2–48.7	.102–.487	63	8.9–68.2	.089–.682
	61–<180 d	164	10.9–47.8	.109–.478	159	14.1–76.0	.141–.760
	0.5–<2 y	596	17.5–69.5	.175–.695	502	16.9–74.0	.169–.740
	2–<6 y	802	22.4–69.0	.224–.690	700	22.4–69.0	.224–.690
	6–<12 y	1006	28.6–74.5	.286–.745	866	29.8–71.4	.298–.714
	12–<18 y	1186	32.5–74.7	.325–.747	1504	39.0–73.6	.390–.736
	≥18 y	270	40.3–74.8	.403–.748	413	42.5–73.2	.425–.732

Specimen Type(s)	1	Whole blood (K$_2$/EDTA anticoagulant)
	2	Whole blood (K$_3$/EDTA anticoagulant)
Reference(s)	1	Brugnara C. Unpublished data.
	2	Children's National Medical Center, Washington, DC. Unpublished data.
Method(s)	1	Bayer ADVIA 120 (Bayer Diagnostics, Tarrytown, NY).
	2	Sysmex XE-2100 (Sysmex Corporation, Kobe, Japan).
Comment(s)	1	Data from outpatients. A computerized approach adapted from the Hoffmann technique was used to remove outliers and obtain the 2.5th and the 97.5th percentiles (1).
	2	Data from outpatients and ER patients, excluding hematology/oncology clinic patients. Data was log transformed and truncated based on a method described by Hoffmann. The top and bottom 25% of data, based on an assumed normal distribution, was discarded (1). The method was modified to allow for computerized handling of the data. The 2.5th and 97.5th percentiles were calculated on the remaining 50% of the data.
		1. Hoffmann RG. Statistics in the practice of medicine. JAMA 1963;185:864–73.

NUCLEATED RED BLOOD CELL COUNT (ABSOLUTE)

Male and Female		
Age	n	× 10^3/μL
1–3 d	50	0.06–1.30
4–30 d	31	0.04–0.11
31–60 d	43	0.03–0.09
61–180 d	50	0.03–0.13
0.5–<2 y	106	0.03–0.12
2–<6 y	92	0.03–0.32
6–<12 y	154	0.03–0.15
12–<18 y	153	0.03–0.13
≥18 y	41	0.03–0.11

Specimen Type(s)	Whole blood (K_3/EDTA anticoagulant)
Reference(s)	Children's National Medical Center, Washington, DC. Unpublished data.
Method(s)	Sysmex XE-2100 (Sysmex Corporation, Kobe, Japan).
Comment(s)	Data from outpatients, nursery patients, and ER patients, excluding hematology/oncology clinic patients. Data was analyzed using SAS statistical software. Outliers were removed according to the Chauvenet's Criterion (1). This criterion uses an exclusion factor based on the number of values in the sample set and is reapplied until no more outliers are excluded. The remaining data was truncated based on a method described by Hoffmann—the top and bottom 10% of data, based on an assumed normal distribution, was discarded (2). The method was modified to allow for computerized handling of the data. The 2.5th and 97.5th percentiles were calculated on the remaining 80% of the data. 1. Young HD. Statistical treatment of experimental data. New York: McGraw-Hill, 1962:76–9,162. 2. Hoffmann RG. Statistics in the practice of medicine. JAMA 1963;185:864–73.

NUCLEATED RED BLOOD CELL COUNT (RELATIVE)

	Male and Female	
Age	n	Per 100 WBC's
1–3 d	52	0.1–8.3
4–30 d	31	0.0–0.0
31–60 d	43	0.0–0.0
61–180 d	89	0.0–0.0
0.5–<2 y	223	0.0–0.0
2–<6 y	256	0.0–0.0
6–<12 y	375	0.0–0.0
12–<18 y	347	0.0–0.0
≥18 y	90	0.0–0.0

Specimen Type(s)	Whole blood (K_3/EDTA anticoagulant)
Reference(s)	Children's National Medical Center, Washington, DC. Unpublished data.
Method(s)	Sysmex XE-2100 (Sysmex Corporation, Kobe, Japan).
Comment(s)	Data from outpatients, nursery patients, and ER patients, excluding hematology/oncology clinic patients. Data was analyzed using SAS statistical software. Outliers were removed according to the Chauvenet's Criterion (1). This criterion uses an exclusion factor based on the number of values in the sample set and is reapplied until no more outliers are excluded. The remaining data was truncated based on a method described by Hoffmann—the top and bottom 10% of data, based on an assumed normal distribution, was discarded (2). The method was modified to allow for computerized handling of the data. The 2.5th and 97.5th percentiles were calculated on the remaining 80% of the data. 1. Young HD. Statistical treatment of experimental data. New York: McGraw-Hill, 1962:76–9,162. 2. Hoffmann RG. Statistics in the practice of medicine. JAMA 1963;185:864–73.

PLATELET COUNT

Test	Age	Male			Female		
		n	× 10³/μL	× 10⁹/L	n	× 10³/μL	× 10⁹/L
1	1–3 d	436	140–238	140–238	363	133–255	133–255
	4–7 d	276	129–271	129–271	327	95–230	95–230
	8–14 d	299	120–297	120–297	357	106–294	106–294
	15–30 d	344	157–406	157–406	414	114–364	114–364
	31–60 d	474	221–471	221–471	411	184–430	184–430
	61–180 d	1184	215–448	215–448	1105	147–423	147–423
	0.5–<2 y	2893	185–399	185–399	2379	211–408	211–408
	2–<6 y	4120	211–370	211–370	3713	190–365	190–365
	6–<12 y	4163	227–350	227–350	3324	219–339	219–339
	12–<18 y	5028	180–299	180–299	5291	192–307	192–307
	≥18 y	2469	177–290	177–290	3837	197–304	197–304
2	0–14 d	56	218–419	218–419	46	144–449	144–449
	15–30 d	51	248–586	248–586	39	279–571	279–571
	31–60 d	115	229–562	229–562	72	331–597	331–597
	61–180 d	280	244–529	244–529	208	247–580	247–580
	0.5–<2 y	1208	206–445	206–445	1058	214–459	214–459
	2–<6 y	1218	202–403	202–403	1099	189–394	189–394
	6–<12 y	1289	206–369	206–369	1084	199–367	199–367
	12–<18 y	1481	175–332	175–332	1773	194–345	194–345
	≥18 y	263	151–304	151–304	458	186–353	186–353

Specimen Type(s)	1	Whole blood (K_2/EDTA anticoagulant)
	2	Whole blood (K_3/EDTA anticoagulant)
Reference(s)	1	Brugnara C. Unpublished data.
	2	Children's National Medical Center, Washington, DC. Unpublished data.
Method(s)	1	Bayer ADVIA 120 (Bayer Diagnostics, Tarrytown, NY).
	2	Sysmex XE-2100 (Sysmex Corporation, Kobe, Japan).
Comment(s)	1	Data from inpatients, excluding hematology/oncology patients. Data was analyzed using SAS statistical software. Outliers were removed according to the Chauvenet's Criterion (1). This criterion uses an exclusion factor based on the number of values in the sample set and is reapplied until no more outliers are excluded. The remaining data was truncated based on a method described by Hoffmann—the top and bottom 25% of data, based on an assumed normal distribution, was discarded (2). The method was modified to allow for computerized handling of the data. The 2.5th and 97.5th percentiles were calculated on the remaining 50% of the data.
	2	Performed on outpatients and ER patients. Hematology/oncology clinic patients were excluded. Data was analyzed using SAS statistical software. Outliers were removed according to the Chauvenet's Criterion (1). This criterion uses an exclusion factor based on the number of values in the sample set and is reapplied until no more outliers are excluded. The remaining data was truncated based on a method described by Hoffmann—the top and bottom 10% of data, based on an assumed normal distribution, was discarded (2). The method was modified to allow for computerized handling of the data. The 2.5th and 97.5th percentiles were calculated on the remaining 80% of the data.
		1. Young HD. Statistical treatment of experimental data. New York: McGraw-Hill, 1962:76–9,162.
		2. Hoffmann RG. Statistics in the practice of medicine. JAMA 1963;185:864–73.

RED CELL COUNT

Test	Age	Male			Female		
		n	$\times 10^6/\mu L$	$\times 10^{12}/L$	n	$\times 10^6/\mu L$	$\times 10^{12}/L$
1	1–3 d	439	3.69–4.75	3.69–4.75	360	3.79–4.76	3.79–4.76
	4–7 d	276	3.98–5.08	3.98–5.08	335	4.05–4.83	4.05–4.83
	8–14 d	304	3.75–4.93	3.75–4.93	355	4.01–4.73	4.01–4.73
	15–30 d	347	3.61–4.46	3.61–4.46	416	3.70–4.59	3.70–4.59
	31–60 d	484	3.24–4.08	3.24–4.08	409	3.55–4.57	3.55–4.57
	61–180 d	1195	3.67–4.61	3.67–4.61	1109	3.63–4.61	3.63–4.61
	0.5–<2 y	2930	3.81–4.74	3.81–4.74	2398	3.83–4.67	3.83–4.67
	2–<6 y	4140	3.92–4.72	3.92–4.72	3729	3.89–4.67	3.89–4.67
	6–<12 y	4183	3.85–4.75	3.85–4.75	3371	3.88–4.72	3.88–4.72
	12–<18 y	5092	3.74–4.93	3.74–4.93	5349	3.79–4.61	3.79–4.61
	≥18 y	2526	3.75–5.07	3.75–5.07	3849	3.75–4.54	3.75–4.54
2	0–14 d	59	4.10–5.55	4.10–5.55	47	4.12–5.74	4.12–5.74
	15–30 d	53	3.16–4.63	3.16–4.63	39	3.32–4.80	3.32–4.80
	31–60 d	111	3.02–4.22	3.02–4.22	72	2.93–3.87	2.93–3.87
	61–180 d	282	3.43–4.80	3.43–4.80	214	3.45–4.75	3.45–4.75
	0.5–<2 y	1210	4.03–5.07	4.03–5.07	1049	3.97–5.01	3.97–5.01
	2–<6 y	1198	3.89–4.97	3.89–4.97	1087	3.84–4.92	3.84–4.92
	6–<12 y	1278	3.96–5.03	3.96–5.03	1084	3.90–4.96	3.90–4.96
	12–<18 y	1467	4.03–5.29	4.03–5.29	1762	3.93–4.90	3.93–4.90
	≥18 y	264	4.18–5.48	4.18–5.48	467	3.70–4.87	3.70–4.87

Specimen Type(s)	1	Whole blood (K_2/EDTA anticoagulant)
	2	Whole blood (K_3/EDTA anticoagulant)
Reference(s)	1	Brugnara C. Unpublished data.
	2	Children's National Medical Center, Washington, DC. Unpublished data.
Method(s)	1	Bayer ADVIA 120 (Bayer Diagnostics, Tarrytown, NY).
	2	Sysmex XE-2100 (Sysmex Corporation, Kobe, Japan).
Comment(s)	1	Data from inpatients, excluding hematology/oncology patients. Data was analyzed using SAS statistical software. Outliers were removed according to the Chauvenet's Criterion (1). This criterion uses an exclusion factor based on the number of values in the sample set and is reapplied until no more outliers are excluded. The remaining data was truncated based on a method described by Hoffmann—the top and bottom 25% of data, based on an assumed normal distribution, was discarded (2). The method was modified to allow for computerized handling of the data. The 2.5th and 97.5th percentiles were calculated on the remaining 50% of the data.
	2	Performed on outpatients and ER patients. Hematology/oncology clinic patients were excluded. Data was analyzed using SAS statistical software. Outliers were removed according to the Chauvenet's Criterion (1). This criterion uses an exclusion factor based on the number of values in the sample set and is reapplied until no more outliers are excluded. The remaining data was truncated based on a method described by Hoffmann—the top and bottom 10% of data, based on an assumed normal distribution, was discarded (2). The method was modified to allow for computerized handling of the data. The 2.5th and 97.5th percentiles were calculated on the remaining 80% of the data.
	1.	Young HD. Statistical treatment of experimental data. New York: McGraw-Hill, 1962:76–9,162.
	2.	Hoffmann RG. Statistics in the practice of medicine. JAMA 1963;185:864–73.

RED CELL DISTRIBUTION WIDTH, CV (RDW)

Test	Age	Male		Female	
		n	%	n	%
1	1–3 d	439	16.3–18.2	364	15.8–17.8
	4–7 d	276	15.0–17.5	333	14.7–16.6
	8–14 d	304	14.9–16.3	353	14.8–16.3
	15–30 d	344	14.6–16.4	413	15.0–16.7
	31–60 d	484	14.7–16.2	402	14.2–15.6
	61–180 d	1183	13.5–15.3	1063	13.6–14.8
	0.5–<2 y	2846	13.6–15.5	2316	13.3–14.8
	2–<6 y	3998	13.2–14.5	3465	13.0–14.2
	6–<12 y	3960	13.0–14.2	3194	12.8–13.9
	12–<18 y	4932	13.0–14.6	5194	12.8–14.4
	≥18 y	2434	13.1–15.2	3787	13.0–14.7
2	0–14 d	66	14.8–17.0	49	14.6–17.3
	15–30 d	56	14.3–16.8	43	14.4–16.2
	31–60 d	111	13.8–16.1	75	13.6–15.8
	61–180 d	298	12.4–15.3	223	12.2–14.3
	0.5–<2 y	1212	12.9–15.6	1064	12.7–15.1
	2–<6 y	1179	12.5–14.9	1074	12.4–14.9
	6–<12 y	1243	12.3–14.1	1090	12.2–14.4
	12–<18 y	1423	12.4–14.5	1737	12.3–14.6
	≥18 y	257	12.3–14.3	155	12.4–15.1

Specimen Type(s)	1	Whole blood (K_2/EDTA anticoagulant)
	2	Whole blood (K_3/EDTA anticoagulant)
Reference(s)	1	Brugnara C. Unpublished data.
	2	Children's National Medical Center, Washington, DC. Unpublished data.
Method(s)	1	Bayer ADVIA 120 (Bayer Diagnostics, Tarrytown, NY).
	2	Sysmex XE-2100 (Sysmex Corporation, Kobe, Japan).
Comment(s)	1	Data from inpatients, excluding hematology/oncology patients. Data was analyzed using SAS statistical software. Outliers were removed according to the Chauvenet's Criterion (1). This criterion uses an exclusion factor based on the number of values in the sample set and is reapplied until no more outliers are excluded. The remaining data was truncated based on a method described by Hoffmann—the top and bottom 25% of data, based on an assumed normal distribution, was discarded (2). The method was modified to allow for computerized handling of the data. The 2.5th and 97.5th percentiles were calculated on the remaining 50% of the data.
	2	Performed on outpatients and ER patients. Hematology/oncology clinic patients were excluded. Data was analyzed using SAS statistical software. Outliers were removed according to the Chauvenet's Criterion (1). This criterion uses an exclusion factor based on the number of values in the sample set and is reapplied until no more outliers are excluded. The remaining data was truncated based on a method described by Hoffmann—the top and bottom 10% of data, based on an assumed normal distribution, was discarded (2). The method was modified to allow for computerized handling of the data. The 2.5th and 97.5th percentiles were calculated on the remaining 80% of the data.
		1. Young HD. Statistical treatment of experimental data. New York: McGraw-Hill, 1962:76–9,162.
		2. Hoffmann RG. Statistics in the practice of medicine. JAMA 1963;185:864–73.

RED CELL DISTRIBUTION WIDTH, SD (RDW)

Age	Male		Female	
	n	fL	n	fL
0–14 d	57	51.0–61.7	47	51.4–65.7
15–30 d	47	46.3–57.3	42	47.2–59.8
31–60 d	110	43.9–52.8	75	43.0–55.0
61–180 d	273	35.3–45.7	210	35.2–45.1
0.5–<2 y	1179	35.3–42.8	1030	34.9–42.4
2–<6 y	1146	35.1–41.7	1042	34.9–42.0
6–<12 y	1224	35.1–41.7	1036	35.5–41.8
12–<18 y	1391	36.7–43.8	1704	37.1–44.2
≥18 y	255	37.8–46.1	449	38.4–47.7

Specimen Type(s)	Whole blood (K_3/EDTA anticoagulant)
Reference(s)	Children's National Medical Center, Washington, DC. Unpublished data.
Method(s)	Sysmex XE-2100 (Sysmex Corporation, Kobe, Japan).
Comment(s)	Performed on the XE-2100 on outpatients and ER patients. Hematology/oncology clinic patients were excluded. Data was analyzed using SAS statistical software. Outliers were removed according to Chauvenet's Criterion (1). This criterion uses an exclusion factor based on the number of values in the sample set and is reapplied until no more outliers are excluded. The remaining data was truncated based on a method described by Hoffmann—the top and bottom 10% of data, based on an assumed normal distribution, was discarded (2). The method was modified to allow for computerized handling of the data. The 2.5th and 97.5th percentiles were calculated on the remaining 80% of the data. 1. Young HD. Statistical treatment of experimental data. New York: McGraw-Hill, 1962:76–9,162. 2. Hoffmann RG. Statistics in the practice of medicine. JAMA 1963;185:864–73.

RETICULOCYTE CELLULAR HEMOGLOBIN CONTENT (CHR)

Age	Male		Female	
	n	pg/cell	n	pg/cell
1 d–<2 y	132	22.5–31.8	104	23.9–30.9
2–<6 y	127	25.1–32.0	92	26.4–32.1
6–<12 y	133	23.6–33.9	116	25.1–33.3
12–<18 y	211	27.0–33.2	221	28.2–33.9
≥18 y	214	30.1–34.6	402	27.1–35.2

Specimen Type(s)	Whole blood (K_2/EDTA anticoagulant)
Reference(s)	Brugnara C. Unpublished data.
Method(s)	Bayer ADVIA 120 (Bayer Diagnostics, Tarrytown, NY).
Comment(s)	Data from outpatients. A computerized approach adapted from the Hoffmann technique was used to remove outliers and obtain the 2.5th and 97.5th percentiles.

RETICULOCYTE COUNT (ABSOLUTE)

Test	Age	Male			Female		
		n	× 10³ /μL	× 10⁹/L	n	× 10³ /μL	× 10⁹/L
1	1–<6 mo	66	37–104	37–104	24	52–120	52–120
	0.5–<2 y	68	29–89	29–89	78	35–92	35–92
	2–<6 y	126	29–80	29–80	92	43–83	43–83
	6–<12 y	129	39–106	39–106	115	37–93	37–93
	12–<18 y	204	39–100	39–100	216	40–102	40–102
	≥18 y	213	43–85	43–85	392	46–102	46–102

Test	Age	Male and Female		
		n	× 10³ /μL	× 10⁹/L
2	1–3 d	55	147.5–216.4	147.5–216.4
	4–30 d	46	51.3–110.4	51.3–110.4
	31–60 d	21	51.8–77.9	51.8–77.9
	61–180 d	31	48.2–88.2	48.2–88.2
	0.5–<2 y	120	43.5–111.1	43.5–111.1
	2–<6 y	111	36.4–68.0	36.4–68.0
	6–<12 y	91	42.4–70.2	42.4–70.2
	12–<18 y	130	41.6–65.1	41.6–65.1
	≥18 y	29	39.1–57.0	39.1–57.0

Specimen Type(s)	1 Whole blood (K_2/EDTA anticoagulant)
	2 Whole blood (K_3/EDTA anticoagulant)
Reference(s)	1 Brugnara C. Unpublished data.
	2 Children's National Medical Center, Washington, DC. Unpublished data.
Method(s)	1 Bayer ADVIA 120 (Bayer Diagnostics, Tarrytown, NY).
	2 Sysmex XE-2100 (Sysmex Corporation, Kobe, Japan).
Comment(s)	1 Data from outpatients. A computerized approach adapted from the Hoffmann technique was used to remove outliers and obtain the 2.5th and 97.5th percentiles.
	2 Data from outpatients and ER patients, excluding hematology/oncology clinic patients. Ages >6 mo excluded ER patients. Data was analyzed using SAS statistical software. Outliers were removed according to the Chauvenet's Criterion (1). This criterion uses an exclusion factor based on the number of values in the sample set and is reapplied until no more outliers are excluded. The remaining data was truncated based on a method described by Hoffmann—the top and bottom 10% of data, based on an assumed normal distribution, was discarded (2). The method was modified to allow for computerized handling of the data. The 2.5th and 97.5th percentiles were calculated on the remaining 80% of the data. 1. Young HD. Statistical treatment of experimental data. New York: McGraw-Hill, 1962:76–9,162. 2. Hoffmann RG. Statistics in the practice of medicine. JAMA 1963;185:864–73.

RETICULOCYTE COUNT (RELATIVE)

Test	Age	Male			Female		
		n	%	Number Fraction	n	%	Number Fraction
1	1–3 d	22	2.2–4.8	.022–.048	26	2.1–3.7	.021–.037
	4–30 d	58	0.4–2.7	.004–.027	29	0.4–2.0	.004–.020
	31–60 d	67	0.9–3.8	.009–.038	34	1.5–3.2	.015–.032
	61–180 d	138	0.9–3.1	.009–.031	98	1.1–2.9	.011–.029
	0.5–<2 y	602	0.8–2.0	.008–.020	439	0.9–2.0	.009–.020
	2–<6 y	966	0.8–2.0	.008–.020	770	0.8–2.1	.008–.021
	6–<12 y	688	0.7–2.2	.007–.022	570	0.8–2.8	.008–.028
	12–18 y	480	0.8–2.2	.008–.022	558	0.8–2.2	.008–.022

Test	Age	Male and Female		
		n	%	Number Fraction
2	1–3 d	65	3.47–5.40	.0347–.0540
	4–30 d	45	1.06–2.37	.0106–.0237
	31–60 d	24	2.12–3.47	.0212–.0347
	61–180 d	32	1.55–2.70	.0155–.0270
	0.5–<2 y	104	0.99–1.82	.0099–.0182
	2–<6 y	99	0.82–1.45	.0082–.0145
	6–<12 y	90	0.98–1.94	.0098–.0194
	12–<18 y	125	0.90–1.49	.0090–.0149
	≥18 y	30	0.86–1.36	.0086–.0136

Specimen Type(s)	1	Whole blood (K_2/EDTA anticoagulant)
	2	Whole blood (K_3/EDTA anticoagulant)
Reference(s)	1	Soldin SJ, Brugnara C, Wong EC, eds. Pediatric reference intervals, 6th ed. Washington, DC: AACC Press, 2007.
	2	Children's National Medical Center, Washington, DC. Unpublished data.
Method(s)	1	Bayer H*3 (Bayer Diagnostics, Tarrytown, NY).
	2	Sysmex XE-2100 (Sysmex Corporation, Kobe, Japan).
Comment(s)	1	Data from outpatients. A computerized approach adapted from the Hoffmann technique was used to remove outliers and obtain the 2.5th and 97.5th percentiles.
	2	Data from outpatients and ER patients excluding hematology/oncology clinic patients. Ages >6 mo did not include ER patient data. Data was analyzed using SAS statistical software. Outliers were removed according to the Chauvenet's Criterion (1) This criterion uses an exclusion factor based on the number of values in the sample set and is reapplied until no more outliers are excluded. The remaining data was truncated based on a method described by Hoffmann—the top and bottom 10% of data, based on an assumed normal distribution, was discarded (2). The method was modified to allow for computerized handling of the data. The 2.5th and 97.5th percentiles were calculated on the remaining 80% of the data. 1. Young HD. Statistical treatment of experimental data. New York: McGraw-Hill, 1962:76–9,162. 2. Hoffmann RG. Statistics in the practice of medicine. JAMA 1963;185:864–73.

RETICULOCYTE HEMOGLOBIN EQUIVALENT (Ret-He)

	Male		Female	
Age	n	pg	n	pg
0–180 d	70	27.6–38.7	51	29.2–37.5
6–<24 mo	54	28.7–35.7	55	30.1–35.7
2–<6 y	55	27.7–37.8	65	29.3–37.3
6–<12 y	56	32.4–37.6	54	30.4–39.7
12–<18 y	56	30.3–40.4	63	29.9–38.4
≥18 y	52	36.0–38.6	52	30.6–40.7

Specimen Type(s)	Whole blood (K_3/EDTA anticoagulant)
Reference(s)	Children's National Medical Center, Washington, DC. Publication pending.
Method(s)	Sysmex XE-2100 (Sysmex Corporation, Kobe, Japan).
Comment(s)	Data from outpatients and ER patients excluding hematology/oncology clinic patient, previously transfused patients and patients with other hemoglobinopathies. Data was truncated based on the method described by Hoffmann—the top and bottom 20–30% of the data, based on an assumed normal distribution was discarded (1). The method was modified to allow for computerized handling of the data. The 2.5th and 97.5th percentiles were calculated on the remaining 40–60% of the data with the exception of ≥18 (male) and 6–<24 mo (female) data where the central 20% was used. 1. Hoffmann RG. Statistics in the practice of medicine. JAMA 1963:185:864–73.

SOLUBLE TRANSFERRIN RECEPTOR (sTfR)

Test	Age	Male and Female		
		n		mg/L
1	0–<1 y	96		0.6–7.8
	≥1–≤5 y	136		0.6–7.4

	Male		Female	
Age	n	mg/L	n	mg/L
>5–≤10 y	51	2.39–9.37	52	2.43–7.08

Male and Female		
Age	n	mg/L
>10–≤20 y	241	0.8–8.5

Test	Age	Male and Female		
		n	nmol/L	mg/L*
2	0.5–<2 y	37	18.2–38.0	1.37–2.85
	2–<6 y	42	14.0–40.7	1.05–3.05
	6–<12 y	39	15.5–36.3	1.16–2.72
	12–<18 y	37	12.9–34.7	0.97–2.60
	≥18 y	28	11.2–30.9	0.84–2.32

Specimen Type(s)	1	Plasma/serum
	2	EDTA plasma
Reference(s)	1	Kulasingam V, Jung BP, Blasutig IM, et al. Pediatric reference intervals for 28 chemistries and immunoassays on the Roche cobas® 6000 analyzer—A CALIPER pilot study. Clin Biochem 2010;43;1045–50.
	2	Kratovil T, Deberardinis J, Gallagher N, Luban NL, Soldin SJ, Wong EC. Age specific reference intervals for soluble transferrin receptor (sTfR). Clin Chim Acta 2007;380:222–4.
Method(s)	1	Roche cobas® 6000 analyzer (Roche Diagnostics Limited, West Sussex, UK).
	2	Quantikine IVD sTfR Immunoassay kit (R&D Systems, Minneapolis, MN).
Comment(s)	1	Approximately 600 outpatient samples from a pediatric population deemed to be metabolically stable were subdivided into five age classes ranging from 0–20 y and further partitioned by gender. Values are 2.5–97.5th percentiles.
	2	Data from outpatients and ER patients (excluding hematology/oncology clinic patients) with normal hemoglobin, hematocrit, and mean corpuscular volume. Data was log transformed prior to truncation based on a method described by Hoffmann (1) in which the top and bottom 2.5% of data, based on an assumed normal distribution, was discarded. 1. Hoffmann RG. Statistics in the practice of medicine. JAMA 1963;185:864–73. *Based on a calculated molecular weight of 75 kDa.

WHITE CELL COUNT

Test	Age	Male n	× 10³/μL	× 10⁹/L	Female n	× 10³/μL	× 10⁹/L
1	1–3 d	423	7.69–13.12	7.69–13.12	356	7.51–15.83	7.51–15.83
	4–7 d	270	6.54–12.32	6.54–12.32	315	5.86–12.23	5.86–12.23
	8–14 d	302	7.66–14.05	7.66–14.05	340	7.46–14.55	7.46–14.55
	15–30 d	346	8.90–16.69	8.90–16.69	414	8.55–15.72	8.55–15.72
	31–60 d	468	8.36–13.66	8.36–13.66	407	7.34–12.32	7.34–12.32
	61–180 d	1163	7.91–13.41	7.91–13.41	1106	6.85–12.84	6.85–12.84
	0.5–<2 y	2778	7.73–13.12	7.73–13.12	2378	7.05–12.98	7.05–12.98
	2–<6 y	4071	5.97–10.49	5.97–10.49	3575	5.98–10.80	5.98–10.80
	6–<12 y	4030	5.69–9.88	5.69–9.88	3305	5.41–9.70	5.41–9.70
	12–<18 y	4987	5.24–9.74	5.24–9.74	5219	5.52–9.29	5.52–9.29
	≥18 y	2432	5.78–10.33	5.78–10.33	3732	5.82–9.32	5.82–9.32
2	0–14 d	56	8.04–15.40	8.04–15.40	44	8.16–14.56	8.16–14.56
	15–30 d	53	7.80–15.91	7.8–15.91	35	8.36–14.42	8.36–14.42
	31–60 d	111	8.14–14.99	8.14–14.99	75	7.05–14.68	7.05–14.68
	61–180 d[a]	67	6.51–13.32	6.51–13.32	44	6.00–13.25	6.00–13.25
	0.5–<2 y[a]	537	5.98–13.51	5.98–13.51	466	6.48–13.02	6.48–13.02
	2–<6 y	1194	5.14–13.38	5.14–13.38	1075	4.86–13.18	4.86–13.18
	6–<12 y	1276	4.31–11.00	4.31–11.00	1068	4.27–11.40	4.27–11.40
	12–<18 y	1454	3.84–9.84	3.84–9.84	1736	4.19–9.43	4.19–9.43
	≥18 y	261	3.91–8.77	3.91–8.77	458	4.37–9.68	4.37–9.68

Specimen Type(s)	1 Whole blood (K_2/EDTA anticoagulant)
	2 Whole blood (K_3/EDTA anticoagulant)
Reference(s)	1 Brugnara C. Unpublished data.
	2 Children's National Medical Center, Washington, DC. Unpublished data.
Method(s)	1 Bayer ADVIA 120 (Bayer Diagnostics, Tarrytown, NY).
	2 Sysmex XE-2100 (Sysmex Corporation, Kobe, Japan).
Comment(s)	1 Data from inpatients, excluding hematology/oncology patients. Data was analyzed using SAS statistical software. Outliers were removed according to the Chauvenet's Criterion (1). This criterion uses an exclusion factor based on the number of values in the sample set and is reapplied until no more outliers are excluded. The remaining data was truncated based on a method described by Hoffmann—the top and bottom 25% of data, based on an assumed normal distribution, was discarded (2). The method was modified to allow for computerized handling of the data. The 2.5th and 97.5th percentiles were calculated on the remaining 50% of the data.
	2 Performed on outpatients and ER patients. Hematology/oncology clinic patients were excluded. Data was analyzed using SAS statistical software. Outliers were removed according to the Chauvenet's Criterion (1). This criterion uses an exclusion factor based on the number of values in the sample set and is reapplied until no more outliers are excluded. The remaining data was truncated based on a method described by Hoffmann—the top and bottom 10% of data, based on an assumed normal distribution, was discarded (2). The method was modified to allow for computerized handling of the data. The 2.5th and 97.5th percentiles were calculated on the remaining 80% of the data.
	[a]Excludes ER specimens.
	1. Young HD. Statistical treatment of experimental data. New York: McGraw-Hill, 1962:76–9,162.
	2. Hoffmann RG. Statistics in the practice of medicine. JAMA 1963;185:864–73.